THIS WAS

WHEAT FARMING

This Was

WHEAT FARMING

A Pictorial History of the Farms and Farmers of the Northwest who grow the Nation's Bread.

by KIRBY BRUMFIELD

A PAGE FROM THE PAST. In 1911, Pendleton wheat rancher W. R. Meiners switched from a stationary threshing machine to this horse-pulled combined-harvester. It was a big event. Wife Jennie in the bonnet and her sister Effie Lange drove the buckboard to the field to see the machine in operation. Sitting next to the teamster on a seemingly precarious seat which extended out over the horses was four-year-old Mervin Meiners.

Courstey, Mervin Meiners.

SUPERIOR PUBLISHING COMPANY-SEATTLE

Library of Congress Catalogue Card Number 68-22356

PRINTED IN THE UNITED STATES OF AMERICA

DEDICATED

to

My Wife Mena

Foreword

There's a great deal of talk these days about a gap between the young generation and the old generation. This so-called "generation gap" was a problem to me long before the term was ever coined and I can thank my parents for that. It's also the reason behind the writing of this book.

Coming from Montana ranches, Mother and Dad passed on to me a love for things of the soil. Then they garnished this interest with fascinating stories of what wheat farming was like in the early 1900's. Homesteads and sod houses, horses and steam engines, grease monkeys and bundle stiffs all sounded so different, so romantic and so fascinating that I frequently found myself wishing I could do something to bridge the generation gap.

It's not that I wanted them to understand my generation better but that I wished I could have been a part of theirs—at least for a little while.

This interest in early-day wheat farming hasn't diminished over the years. I not only blame my parents for that, I thank them for it.

A chance to talk to someone who participated in those early harvests is a treasured moment. Even now, it's almost impossible to keep from confiding what fun it would have been to have participated in those unique threshing seasons.

Out of pity, these weathered harvesters of seasons-gone-by invariably attempt to tone down the appeal of the old-time harvest by telling how lucky those of my generation are to have avoided all the dirt and hard work. They wince a little as they recollect the dusty chaff which always crept inside their clothes and what an irritating nuisance it was — of lifting 145 pound wheat sacks flat from the ground and loading them onto wagons—of the tiresome monotony of pitching wheat into the threshing machine.

For a short while they paint an almost believable picture of why no man in his right mind would ever want to relive those days. It's all a front though, and easily seen through. You know they'd like nothing more than to once again be part of the massive crew of men and horses it took to plant, cut and separate wheat half a century ago.

These farmers of another era betray their true interests as soon as they dredge up the humorous, sometimes dangerous experiences they lived. Always, their eyes take on a new sparkle and their lips break a slight smile as they mentally transport themselves back to those times.

I've often wondered if there wasn't some way to preserve the romance of that period. The answer

came one day in a bookstore at the discovery of Superior Publishing's book titled, "This Was Logging." My immediate thought was that something like that must be done on the history of wheat.

Mr. Al Salisbury, owner-manager of Superior Publishing Company, was warmly receptive to the idea. Interestingly enough, the Washington Association of Wheat Growers had just expressed their interest to him in such a publication.

The resulting book is a credit to the help and cooperation of many fine folks who were generous with their time, information and old photos. Writing a book is a great way to gain a whole new collection of friends. Help came from every direction. The first person to offer assistance was Seab Brogoitti of Helix, Oregon. He invited yours truly to be a speaker at a Helix Chamber of Commerce banquet and requested everyone attending to bring historical farm photos.

C. C. Rittenhouse, also of Helix, has been a big booster. Al Bond, amiable agricultural extension editor from Washington State University, generously opened his files.

Jim Hutcheson, editor of the Walla Walla Union-Bulletin, unearthed several historic features that added much to the storyline. The folks associated with the Oregon Historical Society, the Washington Historical Society, the Idaho Historical Society, The Washington State University Archives and the Oregon State University Archives were invaluable.

Photographer A. M. "Bert" Kendrick of Ritzville, Washington was a prime source of help. Mrs. Harriet Moore of Corvallis, Oregon could always be counted on for her picture-sleuthing talents.

Members of the Western Steam Fiends Association were valuable historians. Al Herman, Jeff Richardson and Ed Paulson were Steam Fiends called on the most. Ted Worral, Carl Kirsch, Clarence E. Mitcham and Rodney Pitts also were a big assist.

Dr. C. S. Holton, U.S.D.A. plant pathologist in charge of the W.S.U. Smut Control Laboratory, added much worthwhile information with his material on the history of smut in Northwest wheat.

Frank Ballard, former director of the Oregon State University Agricultural Extension Service, and Charlie Smith, former assistant director, were both brimming with knowledge on the subject.

Certainly a key figure behind a large number of the photos in this book has been Theodore Johnston of the Sherman County Historical Society. His interest and concern has resulted in Sherman County having one of the most extensive collections of early-day harvest photos in the Northwest.

Thomas W. Thompson, former county agent in Sherman County, led me to Mr. Johnson who

Courtesy, Dick Jennings

dug out numerous photos from his own files. Two other Sherman County personalities who provided stimulation and information were Giles French and Gordon Hilderbrand.

Marion Weatherford, historian and wheat rancher, brought out pictures and little known facts that added greatly to the book. Joe Johnson, animal science specialist, W.S.U. Agricultural Extension Service, shared the vision. So did Cal Crandall, executive director of the Oregon Dairy Products Commission. The Charles Freeburg collection of old photos preserved by Mrs. Freeburg contributed much.

Dr. Norman Goetze, agronomy specialist, O.S.U. Agricultural Extension Service, made the job of digging out information much easier as he was always available with technical facts vital to the study. F. Hal Higgins, curator of his well known collection of facts and photos on early steam and gas machinery from the University of California at Davis, was a most appreciated aid.

The Washington Association of Wheat Growers not only supplied moral support but also solicited their members for pictures that added immensely to the pictorial history. The Oregon Wheat Growers' League and the Oregon Wheat Commission supplied much in background information.

Al Salisbury of Superior Publishing has been a helping hand all along.

This is just a portion of the long list of people that made contributions. My sincerest thanks goes to each of them.

This book, of course, would have been impossible if those turn-of-the-century wheat growing pioneers had not dared to stake their future to the golden grain—wheat. To them we owe the greatest debt. This is their story.

The pictures and text deal primarily with early-day wheat ranching in Oregon, Washington and Idaho. This area comprises the bulk of the soft, white wheat region. Most other vicinities grow hard, red wheat types.

The bulk of the photos chosen range in age from approximately 50 to 70 years. The quality of most of these old prints is amazingly good. Occasionally, harvest pictures slightly faded or worn were chosen because of the valuable story they told.

Writing and compiling this book has been an interesting journey. It has served as my own personal bridge across the huge gap separating this generation from an earlier one. It has given me some reality as to what life on a wheat ranch was really like way back then. "This Was Wheat Farming" will be a success in the author's mind if it serves as the same kind of bridge for every reader.

Kirby Brumfield

Courtesy, Washington State University Archives

Table of Contents

                                                                    Page

CHAPTER I   WHEAT'S HISTORY IS MANKIND'S HISTORY.....   11

                Bread   15

                Mechanical reapers ushers in new era.   15

CHAPTER II   WHEAT IN THE NORTHWEST   17

                Flour mills   25

                Acquiring land   33

                Life was tough   34

                Rained out   35

                Wheat is king   36

CHAPTER III   THE LORE OF THE HARVEST.   51

                Binding and threshing   51

                Heading and threshing   56

                Threshing   61

                Crewmen   61

                Women in the harvest.   67

                Sweep power   82

                Steam power   88

CHAPTER IV   HEROES OF THE HARVEST   109

CHAPTER V   COMBINED HARVESTER   119

                32 horsepower   120

CHAPTER VI   MODERN HARVEST   157

CHAPTER VII   WHEAT GROWERS ASSOCIATION   163

                Tractor-horse cost comparisons between two groups
                   of farms of the same size.   164

                Research and rescue   171

CHAPTER VIII   SMUT   175

                Smut history.   175

                Smut hits all-time high   176

CHAPTER IX   WHEAT—PAST, PRESENT AND FUTURE.   185

Chapter I

Wheat's History Is Mankind's History...

REACHING INTO A GRAIN BIN and scooping up wheat is almost like grabbing a fistful of history. No one knows exactly where or when the wheat plant originated but there's evidence aplenty that it has been around a long time. Archeologists can prove this as they dig through the rubble of generations past. In the process of spading and probing, scientists from the University of Chicago in 1948 uncovered a long-forgotten village in Iraq. By their calculations it had been a thriving community some 6700 years before.

As they sifted through the dirt and dust, their trained eyes spotted two different kinds of wheat very similar to the varieties we grow today. On other occasions specialists sleuthing into the past have found wheat in Egyptian mummy cases. Scientific researchers feel that even then wheat was an ancient food.

The development of grain agriculture in prehistoric times was certainly one of the great turning points in human history. There's a glamour attached to hunting down wild animals, but from a hard-nosed, analytical standpoint, there is a more practical way to provide food.

The discovery that the seeds of the plant we call wheat could be eaten, that they could be stored to serve as food during winter and as seed for a new crop in the spring no doubt had a settling influence on early-day wanderers.

In many unexpected ways the story of wheat is the story of our civilization. The thought of private property likely developed when early man built a shelter so he could stay near his wheat and claim his crop. Villages developed where land was plentiful as people joined together for mutual protection.

No longer wandering aimlessly from place to place, man spent more time thinking of ways to improve his crops, his shelter and his way of life. He began to watch the stars since wheat must be planted at the proper season. In time astronomy led to calendars and mathematics.

Farmers, even then, couldn't live in a vacuum. They needed to trade ideas and understand each other. This led to the development of language. To simplify communication, symbols were devised which marked the beginning of writing.

The first large administrative machine resembling a modern national government was also related to wheat. Egypt was one of the world's first wheatlands. Its far-flung network of irrigation canals required supervision and organization.

Mankind's complete reliance on muscle-power came to an end with the development of the mill-wheel. Stey by step, the elements of modern society were fashioned because of a wheat-structured economy. It's hardly stretching a point to say the seeds of wheat became the seeds of civilization.

SICKLE WAS SLOW. The earliest known harvesting implement was a sickle. It may have been better than having no harvesting tool at all but even then it was a slow and backbreaking way to harvest the wheat crop. Cutting a half acre a day was the most a good man could accomplish.

Courtesy, International Harvester Company

Today, wheat is the world's number one farm crop. One acre of every seven of the world's working farm land is producing wheat. With its acreage adding up to nearly as much as rice and corn together, it's easy to see that it occupies more land than any other food crop.

Wheat's slender stalks can be found almost anywhere man is found. That covers a wide area extending from the poles to the equator, and from sea level to the 11,000-foot slopes of the Himalayan mountains. The number of wheat strains runs into the thousands—approximately 15,000.

At least one of these varieties will grow successfully in the hottest or coldest civilized countries. The wettest and driest places on earth can also lay claim to the golden grain. It has even been said that every moment of the year, somewhere on earth, farmers are reaping a golden harvest of wheat.

Single out an individual wheat seed and there's room to wonder how such a tiny thing could play so important a role in man's past—and his future. This little dynamo of nourishment is barely a quarter of an inch long.

Occasionally, a curious mind will ask how many of the little nut-like berries are in a pound. The answer of 13,000 could be categorized as useless information except that it does point up the tininess of the grains.

The March 1914 issue of American Thresherman

FROM SICKLE TO SCYTHE. A man was able to step up his harvest pace a bit once the scythe came into use. The attached cradle caught the cut grain. After several strokes, the cradle would be sufficiently full of wheat to be dumped. This was done by merely bringing the implement back and up to let the wheat slide from the cradle to a neat pile on the ground. Following along in the cutting paths were men who bound the bunches of grain into sheaves. They tied the sheaves or bundles with stalks of wheat. Using a scythe and cradle a man could cut about two acres a day.

Courtesy, International Harvester Company

featured a short piece designed to stimulate mental gymnasts. It started with an interesting supposition.

"Suppose," it said, "all the wheat in the world were to be suddenly destroyed with the exception of one grain. If all the wheat produced each year were planted, how long would it take starting with this one grain to again equal the present world's production? The world's production of wheat is at present about 4,000,000,000 bushels a year.

"The following table shows that with only an average grain to start from, it would take only fifteen years to again supply the world with wheat. The ratio of success is based on the average production of fourteen bushels per acre in the United States.

"The average rate of seeding is about one and one-fourth bushels an acre, the increase is about eleven times or one bushel of seed will produce eleven bushels of grain.

Starting with one grain . . . . 0.0012 oz.

BIRTHPLACE OF REAPER. The forge shop on the McCormick farm in Walnut Grove, Rockbridge County, Va., as it appeared when Cyrus Hall McCormick invented the reaper. The earliest stages of the reaper can be seen in the foreground. The forge shop still stands.

Courtesy, International Harvester Company

1 year	0.0132 oz.	
2 years	0.1452 oz.	
3 years	1.5972 oz.	
4 years	1 lb. 1.6 oz.	
5 years	12 lb. 1.3 oz.	
6 years	132 lb. 13.8 oz.	
7 years	24 bu. 53 lb.	
8 years	273 bu. 43 lbs.	
9 years	3,014 bu.	
10 years	33,154 bu.	
11 years	364,694 bu.	

12 years . . . . 4,011,634 bu.
13 years . . . . 44,127,974 bu.
14 years . . . . 485,407,714 bu.
15 years . . . . 5,339,484,854 bu.

"The initial weight of 0.0012 oz. for a single grain is the average of ten thousand grains. One thousand grains were selected at random from each of the common varieties of wheat."

This, of course, assumes that every grain grows and reaches maturity to reproduce itself. You can bet that if something really did happen to the bulk of the world's wheat supply, man would jealously guard what was left, knowing how important it was to his own survival.

One of the key reasons that man wasted little time in adopting wheat as his chief cereal plant is simple. It proved to be an absolute dynamo of nourishment. Wheat seeds are fairly bursting with the nutritious qualities that lead to healthy living.

Around the outside of the kernel are the dark-hued germ and bran. They are loaded with thiamine and riboflavin.

Remove these and you expose the creamy white endosperm. This makes up the bulk of the kernel and consists of starch granules scattered through a mass of protein.

The endosperm has abundant protein. The American diet relies heavily on wheat's protein. In fact it provides one-fourth of all the protein in our menus, plus 40 percent of the thiamine.

Surprisingly, wheat often goes unnoticed in much of our daily diet. It's a part of so many foods we take it for granted. A homemaker whipping together a cake-mix isn't giving much thought to what part wheat is playing in the process. Crunching a pizza pie crust or tackling a breakfast roll seldom conjures up visions of waving fields of golden wheat.

BREAD

The most popular wheat food quite obviously is bread. The history of bread goes back to the earliest of times. A form of bread was found in the remains of a stone-age village of Swiss lake dwellers. Bronze tablets dating from the ninth century before Christ picture the grinding of wheat and making of bread in Assyria.

Herodotus, a Greek historian, wrote of Egyptian bread-baking in the fifth century. Tombs along the Nile river contain murals which depict the planting of wheat, the harvest, the grinding of flour and bread-making. Egypt is recognized as the country where leavened bread was started.

Leavening is generally thought to have been discovered purely by accident. As historians reconstruct the story, it was about 5000 years ago that an Egyptian baker made one of man's first chemical discoveries.

It could easily have happened that in preparing a batch of dough a busy baker left some standing too long in the air. With conditions just right, microscopic air-borne yeast cells fell on the moist dough and fermentation began.

What a surprise when he baked it. Instead of the flat bread he had been used to, the Egyptian found the product swollen to a fluffy loaf several times its original size and full of tiny air pockets. He had accidentally discovered raised or leavened bread.

In this capsuled look at wheat's earliest beginning, it is quickly obvious that man has leaned heavily on the wheat plant down through the ages. That's why it can be said a person with the right flicker of imagination could see the story of man in a grain of wheat.

MECHANICAL REAPER USHERS IN NEW ERA

Once the value of wheat was obvious to man, it became a definite part of his life. Wherever civilized man ventured he was sure to take wheat with him. It seemed to virtually spring up in his footsteps. It follows then that wheat would top the popularity list of cereal grains grown by early American farmers as the 19th century got under way.

Unfortunately, the men of the soil were stuck with a hand-labor approach to farming that seems extremely primitive today. Farm mechanization was almost non-existent. Farming methods had progressed little from the days of the Egyptians and Israelites. Farmers were harvesting wheat with the sickle and scythe. As soon as the wheat ripened to its familiar golden hue a line of men cut their way through the standing crop. Others followed in the cutting path or swath to collect the bunches of grain and bind them into sheaves.

A step forward came when a wooden framework called a cradle was added to the scythe. The curving fingers caught the grain as it was cut, leaving it bunched and easier to gather for tying into bundles. It was an improvement, but far from any kind of mechanical breakthrough. A strong man trained in its use could cut only two acres of wheat a day.

Separating the grain kernels from the stalks was still done with a flail or by having cattle tramp over the wheat heads again and again as the animals moved about in a circle. It's no wonder that 90 percent of the population lived on farms. It took that many to provide the food for the infant nation.

The bottleneck in mechanical progress on the farm was broken in 1831 when Cyrus Hall McCormick demonstrated a reaper. McCormick was born February 15, 1809 in the valley of Walnut Grove, Rockbridge county, Va. Even as a young boy he displayed a knack for things mechanical. When only 15, he invented a lightweight grain cradle. With it he could keep up with the older men reaping the wheat.

It was the father, Robert McCormick, who first got Cyrus to thinking of ways to ease the wheat harvest. The eldest McCormick tried several different approaches. He was a versatile man and brought a number of different talents to the task. Besides farming, he was weaver, and operated a sawmill, a smelter

HISTORIC EVENT. The first public trial of the revolutionary reaper took place in July, 1831. The historic field was near Steele's Tavern, close to Walnut Grove. Doubtful neighbors were watching as the machine cut its first swath. While young McCormick walked behind his machine, a negro servant named Jo Anderson raked the platform clear of the cut grain.

Courtesy, International Harvester Company

and a blacksmith shop. The father made a last attempt to invent a mechanical reaper in 1831. Meanwhile, Cyrus, who was only 22, studied his father's efforts, and then conceived his own new principles and in six weeks time produced a machine which cut grain successfully.

It was July, 1831 that the first public trial of the reaper took place. The historic spot was a field near Steele's Tavern, close to Walnut Grove. Dubious neighbors were watching as the young inventor walked behind his machine.

The first reaper was a two-wheeled, horse-drawn assemblage. As it moved forward it brought a line of reciprocating, scissor-like blades against the grain stalks and clipped them close to the ground. Assisting was a rotating paddle wheel which pushed the wheat against the approaching cutting blades and then back onto a platform as it was cut. A Negro servant named Jo Anderson raked the platform clear of the cut grain.

The curious neighbors didn't realize they were witnessing the first step in mechanized harvesting, making the jump from manpower to horsepower. The McCormick reaper was a spur to the American industrial revolution and was the first step toward farm mechanization.

The succeeding chapters in this book will deal with the history and the development of the wheat industry in the Pacific Northwest.

Chapter II

Wheat In The Northwest...

WHEAT HADN'T BEEN INTRODUCED to that area we now call Oregon and Washington as the year 1800 rolled across the calendar. The reason being that the white man hadn't yet trekked the plains to settle the Pacific region. His coming wasn't far off, though. Lewis and Clark reached the Far West on their historic hike in 1805.

Benjamin Bonneville is credited with taking the first wagons through the South Pass in the 1830's. Settlers began flowing over the Oregon Trail about 1841. By 1843, there were so many that a provisional government was organized.

The Oregon Territory was created in 1848 and included the present states of Oregon, Washington and parts of Idaho and Montana. The rugged pioneers wasted little time moving in on the virgin land.

From the standpoint of wheat, timing was perfect. Hand harvesting of the golden bounty would soon be a thing of the past. Although some of the sickle and scythe work was done by the first arrivals, McCormick's reaper pointed to better things ahead. The floodgate was open. Minds had been stimulated by the reaper. McCormick and others began pouring new machines onto the farm market in the mid-1880's.

This rush to farm mechanization was ideal for the development of a huge wheat industry in the Northwest. The early-day farmer would now be able to handle the vast acreages he was soon to cultivate.

One of the foremost historians on the subject of wheat in the Northwest was E. R. Jackman, crops specialist for the Agricultural Extension Service of Oregon State University. He summed up the regional history of the vital grain in a story for the Oregon Farmer magazine, August 6, 1959.

It was titled, "WHEAT—OUR NO. 1 CROP SINCE PIONEER DAYS."

His gifted pen related much about those early beginnings as he wrote the following:

"The Illinois family in 1840, seized with 'Oregon Fever,' took a year or two to prepare for the long overland journey. Clothes were mended and strengthened. Lists were made and remade. Cherished house plants and unneeded clothing were given away. At last a farm auction was held and a thousand things—'articles too numerous to mention'—were sold to the highest bidder, whose bids were often accompanied by sobs from the farm women who could scarcely bear to see a cherished possession pass into alien hands.

"But at last, in the spring of 1842, the family was ready to start. They had one or more strong

17

WHEAT HARVEST
CHANGES WITH FAMILY
Each generation of the Prince
family has harvested wheat a
different way. In 1906, Henry
A. Prince, shown running the
header in the bottom center of
the photo, was surrounded by
horses, men and steam. The farm
was near Thornton, Washington.
Powering the operation was a
25 HP Buffalo Pitts twin cylinder
steam engine.

Courtesy, Bill Walters

wagons pulled by strong, patient, sturdy oxen. Oxen, rather than horses, because they did not stampede nor stray so readily; Indians didn't try to steal them; their feet stood the long trail better; and, in the case of extreme hunger, they were better for eating.

"Stowed neatly into the wagon, with a stern eye for space, were all the possessions needed to make a new home in the wilderness. A few tools, a few cooking utensils, clothes for different kinds of weather, and the bedding, always including a feather-bed, with the load arranged so that the bed could be spread out on top of the load to accommodate a sick owner—if 'trail fever' struck him. The diaries that year and next had daily entries, 'Passed seven new graves today,' 'Counted only four new graves today.'

"But there was one thing in every wagon, the one thing more essential than all others—food. And it was three-fourths flour and wheat. There was bacon, beans, salt pork, coffee and sugar. But it was flour and wheat that was hoarded and kept, even if the hardships made it necessary to throw away clothes, bedding, and pots and pans. Flour to prepare in some form every day. Wheat to carry on one's back if necessary, and eat kernel by kernel, as the last resort.

"So wheat was foremost in the thoughts of every immigrant. He ate it on the way and kept some, if he could, to go into the ground first when he finally arrived. Their talk was on wheat, and most of the diaries carefully noted the prices charged for flour at the remote trading posts that sprang up along the

TRACTOR-POWERED COM-
BINE. By 1940, wheat harvesting on the Prince ranch had
changed drastically. A cat and
combine did the job. The crew
had now dwindled to five.
Henry's son, Burdette, was the
combine man to the left of the
picture. Earl Parrish was driving
the Caterpillar tractor.

Courtesty, Bill Walters

TWO PRINCES DO JOB IN '60s. Grandsons, Eugene and Hubert Prince, each drive a combine in the present-day harvest. Each generation of the Prince family has responded to rapid changes in wheat harvest techniques that have occurred in the past 60 years. Town in background is Thornton, Washington.

Courtesy, Burdette Prince

route. At Fort Hall, flour was $20 a hundred; near La Grande, $40 a hundred.

"The place of wheat in the economy of the new land was striking. In 1845 an Oregon territorial* law was passed setting up how debts might be paid, because a man might be considered prosperous, but he might not have any money. The law stated that for payment of taxes and satisfaction of court judgment, the following were legal tender: Gold, silver, treasury warrants, approved orders on solvent merchants, and wheat.

"The wheat then was mainly in the Willamette valley. Right from the first the settlers knew that Eastern Oregon would grow it, but there wasn't much local market for it in Eastern Oregon and it was too expensive to ship down the river or over the Barlow trail. Freight charges in 1865 by river and portage, from Portland to Umatilla, were $45 per ton, which prohibited much commercial agriculture.

"By 1863 this condition began to focus attention upon railroads and there were many local promoters with grand schemes. A road was building to San Francisco and it reached there in 1869. It was almost intolerable to Oregonians that California, a Johnny-come-lately so far as they were concerned, would have a railroad when Oregon had none. This situation would force Oregon to send to California for all of their eastern supplies, and all mail would have to go that way.

"So, in 1863, a public subscription provided money to make a thorough feasibility survey for a railroad to connect Portland with the new railroad at some point in California. The interesting thing is that nearly all of the subscriptions were in wheat!

"In the meantime, the gold rush had occurred. The mines were pretty much concentrated in Northern California and there was also considerable mining in Southern Oregon. For years these mining settlements were provisioned by farmers in the Rogue River and Umpqua Valleys and also by farmers near Portland. Ships came up the Umpqua as far as Scottsburg, bringing supplies from Portland and from the east.

"From there, long trains of pack mules set out for the Oregon and California mines, but on their return trips they carried loads of flour from the mills in the Rogue River valley. The ships took this flour to California points, where it was again packed into the mines. The going price in the California mining towns was $25 a barrel for Oregon flour.

"So the growing number of flour mills in the interior of Oregon helped to make life tolerable in the mushroom mining towns in California.

"Gold also stimulated wheat growing in Eastern Oregon. Held back by lack of market at first. The gold strikes on the John Day, the Burnt River, and on Powder River and Eagle Creek in Oregon, and those in Idaho at the same time, quickly built small cities at such places as Canyon City, Granite, Auburn, and Silver City. So finally there was a wheat-consuming public and in 1870 three acres of wheat were first grown near where Weston is now.

"But by 1890 wheat was a bonanza crop and in 1893 Umatilla county grew 4,500,000 bushels. Of course, the railroad had reached Portland in 1883 and the wheat began to flow to the ports of all nations.

"Wheat in those days was made into coarse flour in the home, much as coffee was ground by the housewife; but wheat for commerce needs commercial mills somewhere along the line.

"Jedediah Smith, David Jackson and William Sublette, mountain men, visited Vancouver in 1825. This busy fort was owned and run as a British outpost by the Hudson's Bay Company, with Dr. John McLoughlin as chief factor. The Frontier fur trappers used their eyes and later reported in detail. They said that the British raised that year 700 bushels of wheat 'full and plump, making good flour,' and that 80 bushels of it were saved for seed in the spring. They noted 'a grist mill, worked by hand, but intended to work by water.' This was Oregon's first mill. Hand operation was too slow and in 1830 a grist mill was built to be run by oxen.

"Finally, William Cannon, a carpenter and wheelwright, strayed into Vancouver and Dr. McLoughlin commissioned him to build a water power mill at Vancouver. In 1834, Cannon went to Champoeg, where he lived while he built a mill there for new settlers on French Prairie. This mill was owned by Webley Hauxhurst, a squawman, and was the first mill in the Willamette valley.

"Dr. McLoughlin, writing in 1837, said that the settlers near French Prairie grew 5,000 bushels of wheat that year for their new mill. In 1839 he reported the sale of 3,000 bushels of wheat to the Russians at Kamchatka and this trade continued until the Oregon country was ceded to the United States.

"Then came the big wave of immigration, starting in 1842. Mills sprang up all over and nearly every town in western Oregon grew up because it was a mill site and had water power.

"Salem was a town because Mill Creek furnished the power for a mill built by Jason Lee. By 1844 Oregon City had four mills. The owners were Dr. John McLoughlin, John Force, F. N. Blanchet and Henry Bixton.

"Dr. Doney Baker brought the first pair of mill stones to Oregon and he built a mill at Oakland about 1851. Jason Lee brought the first cradles around the horn in 1840. They were of three kinds: The Turkey wing, with nearly straight handles; the muley, with curved handle, and the grapevine with a much crooked handle.

"Other early-day mills were at Brownsville, Silver-

---

*Oregon had no real territorial status until three years later, but the settlers organized anyhow and made their own laws.

20

TETON BACKDROP. The beautiful Tetons in the background stand guard over a field of ripening wheat in eastern Idaho.

Courtesy, Idaho Historical Society

THE NOBLE WORK HORSE. Few people realize the contribution horses and mules made to the wheat harvest years ago. The horse and the mule supplied the muscle needed in cultivating, planting and harvesting during those early days of the Pacific Northwest wheat industry. Many claim the most colorful segment of Northwest wheat history was that period featuring the combined-harvesters pulled by 32 head of horses or mules. No other part of the country used these huge teams to such an extent. Midwesterners found it difficult to believe such operations existed.

Courtesy, Washington State Historical Society

ton, Scotts Mills, Elkton, Roseburg, Oakland, Myrtle Creek, Canyonville, Yoncalla, Springfield, Harrisburg, Kings Valley, Syracuse (near Jefferson).

"The provisional legislature concerned itself with these mills. There were eight of them at the time and evidently one or more of them was not above trying to grind out some extra money with the grist, because one of Oregon's first laws regulated the allowable share of the miller. At this time there was very little money, so most farmers took their wheat to the mill, had it ground, and the mill owner took part of the flour as his toll.

"This early law says that the miller is entitled to 'one-eighth part of all wheat, rye, or other grains, ground and bolted . . . and one-tenth of all . . . grain ground or chopped only . . . and one-seventh part of all Indian corn ground in said mill.'

"The act goes on, 'the miller is held accountable for the safekeeping of all grains . . . and for the bag or bags, cask or casks (in which the grains were delivered), provided that the bag or bags be distinctly marked with the name, mark or brand of the owner thereof.'

"As noted, mills did not come to Eastern Oregon until later, with the exception of a small mill built and operated by Marcus Whitman on Mill Creek at the mission near where Walla Walla, Washington is now.

"Another Eastern Oregon mill of record was at the Warm Springs Indian Agency, but this was a small affair. James A. Allen built a good mill at Prineville in 1875, where it served the entire interior country for years.

"Before the mills were common, flour was brought

22

HORSES OR MULES—WHICH WERE BEST? There was never an end to that argument. There were advantages and disadvantages both ways. Some horsemen claimed mules were quicker to kick their handlers than were horses. There may have been some truth to that considering the defensive comments which were part of a letter written by mule supporter B. F. Swaggart in a letter to the Oregon Agriculturist magazine. It was printed in the November 15, 1910 issue. He had one of the world's most unusual explanations for the occasional strange action of a kicking mule. This is what he had to say. "It is true the nature of the mule differs from that of the horse. It is equally true that in order to handle the mule properly, one must make a practical study of the natural tendencies and characteristics of this animal. In order to instill trust and confidence in the mule, your keynote must be attuned to kindness in handling him. The mule never kicks except at a time when he fears something. Many a person believes that the young mule must be abused and beaten into submission, but such method of handling is wrong and despicable.

"Take for instance, a mule that is handled by a person possessed of fear—what would be the result? The person's magnetism on account of being tinged with fear, is of negative quality, which is attracted by the mule's negative magnetism, and, according to the laws of Occultism, as two negatives or two positive qualities, in contact with each other, cause discord and inharmony, so the mule's negative magnetism repels the person's negative magnetism, and arouses fear in the person, who, in turn causes fear in the mule. The mule, because he fears something, is then liable to kick the person." This may have been good basic knowledge to know but it didn't make a kick hurt any less.

The one year old mule colt in the picture was held by George N. Crosfield of Wasco, Oregon.

Courtesy, Sherman County Historical Society

to the fur traders by ships from Chile. These ships took on hides in California and furs in Oregon.

"But with the discovery of gold in California, the Oregon flour mills entered a 20-year period of prosperity. The mines would take all the flour they could send and it was shipped by water from Portland and Scottsburg (Douglas County) and by mule pack train.

"Markets soon extended and in 1867 Oregon flour was reported as the highest priced and best flour on the New York market. The writer of this report claimed irritably that 'since the boats came by way of California, this flour usually sold as California flour.'

"In 1866 a writer said that in that year $149,065 worth of flour moved from Portland for California, where it sold for $5 a barrel. But at the mines in California markets, it brought up to $25 a barrel.

"The earliest recorded export was a report by

Avery Sylvester, in command of a ship from Boston. He tells of shipping flour and other products in 1845 from Astoria to Boston by way of the Hawaiian Islands. Trade with Hawaii was rather common. Islanders even made up part of the crews of the fur brigades, notably Peter Skene Ogden; and Oregon's Owyhee River was named after the islands, since Owyhee is nearer to the island pronunciation than our word Hawaii.

"In the 1860's wheat furnished the main cargo for many boats operating on the Willamette. In December, 1861, a tragedy occurred. The crop of wheat had been excellent and every warehouse from Eugene to Portland bulged with it. Then the rains started, and the water rose and rose. It carried away some mills and some warehouses, including the large mills at Oregon City. And it swirled into the others, the wet wheat

DINNERTIME AT THE PORTABLE CAFETERIA. Having a colt didn't exempt a mare from work, at least not for long. It was not uncommon to see a mother harnessed in with the other horses while her young offspring followed along nearby.

Courtesy, Oregon State University Archives

swelled and the warehouses burst. The remaining wheat heated and sprouted and the river boating was a dismal, unprofitable business that year.

"The new Oregon country, of course, desperately needed money. Settlers bought clothes, coffee, furniture, spices, household and farm tools and utensils—and money for these was sent to the east. They mostly started with nothing, and there was an immense need for stoves, for example. Flour, guns, harness—everything took money. So wheat furnished the money.

"The 'Commercial Review' (Portland) says that in 1868 'the little bark Helen Angier is credited with carrying the first shipment of wheat overseas from Portland.'

"This was evidently successful because within two years, wheat shipments were an established fact. The bark Sallie Brown came from New York around the 'Horn' in 170 days and returned with a full cargo of wheat and flour. In July, 1870, Rober Meyer & Co. loaded the German ship Herman Dokter with 11,468 centals of wheat. This ship was from China, and illustrates a common route. Atlantic coast ships carried goods to China and returned via Portland and San Francisco. In that year, 12 vessels cleared from Portland with 242,579 bushels of wheat.

"The Willamette Valley was then providing the oil for Oregon wheels of commerce by growing wheat. But by 1870, the valley was too crowded, too conservative, for the restless spirits. Gold had lured some, but the rich strikes were getting scarce, and many miners in what is now Baker and Grant counties found that long days with a shovel dulled the glitter of gold. These men began to stand up, look around them, and they liked what they saw. They promptly homesteaded.

"So the settlement of eastern Oregon grew rapidly from 1870 on. Many came, again in covered wagons, but this time the wheat rolled from west to east. There are still a few of these old wagons parked under trees on eastern Oregon ranches; one, for example on the Locey Ranch near Ironside in Malheur County.

"But, as noted, freight charges down the rivers were high and the early comers mainly grew cattle and sheep because animals could walk to market. From 1870 to 1883 wheat pressure built up behind the Cascades and some wheat trickled through to the waiting export markets by barge, portage, wagon, river boat, and mule back. In 1883 the Union Pacific railroad reached Portland and dammed-up wheat burst through.

"In the meantime, the mining towns were dying

24

of malnutrition, and the profitable Willamette Valley and Southern Oregon trade with the mines slowly petered out so that by 1890 there wasn't much left of this business. Wheat never again dominated the western Oregon business life or farm life.

"Wheat is still grown in rotation in Western Oregon, of course, but the acreage now is roughly 10 per cent of the state total, whereas 60 years ago half of our wheat acres were west of the Cascades.

"So Oregon's export wheat business started in Western Oregon, built our towns and farms, but the eagerly-sought railroad came and doomed the business. The railroad rails led straight to the larger fields of Eastern Oregon and Washington where wheat could be grown more cheaply."

Mr. E. R. Jackman thus summarizes the story of wheat in the Oregon country. His references to wheat getting a toe hold west of the mountains first and then moving east across the Cascades is also the way it happened in Washington—Vancouver first, then to Walla Walla, into Idaho and to all points that now grow the adaptable grain.

FLOUR MILLS

In those early days, it was logical that flour mills would sprout up in practically every community where wheat was grown. A detailed look at the early beginnings of wheat farming and the resulting flour mills in the northwest was published in the Walla Walla Union-Bulletin February 13, 1938. It serves as an informative supplement to Jackson's story. The information was supplied in an interview with J. J. Mulvey, pioneer miller for the Collins Flour Mills at Pendleton. He secured many of the details in the course of preparing a historical treatise on early-day mills for publication in the Northwestern Miller &

BLACKSMITH. Pioneer blacksmith Conrad Sittner was an invaluable help to farmers of the Ritzville, Washington area from the turn of the century until about 1938. An artist with his tools, Sittner fixed wagons, set tires, ground plowshares, made clevises and hitches and could do just about anything a farmer asked him to when it came to fixing or making machinery.

Courtesy, A. M. Kendrick

WATERING TIME. You either had a strong arm or you developed one when it came time to water this many horses. The water wasn't piped to this trough, it was pumped to it. The sturdy young farmhand glancing over his shoulder at the camera had his right hand on the pump handle. Soon after the shot was taken he went back to the tiresome but necessary chore of providing thirst-quenching water for the horses and mules just in from summer-fallow work in the fields. The farm was located near Pendleton, Oregon. *Courtesy, Buz Howdyshell*

American Baker. For the newspaper, Mulvey summarized his material as follows:

"As chief factor for Hudson's Bay Company at Fort Vancouver, Dr. John McLoughlin realized the pressing need of wheat production in the Oregon country as early as 1825. That spring he caused to be brought overland from York factory, at the mouth of Hayes River, Hudson Bay, one bushel each of wheat, oats, barley and corn. This was carefully seeded and reseeded until, by 1828, the first bushel had been multiplied sufficiently to suspend importation of flour for the Oregon country from England.

"The first mill, so far as I can learn, erected and operated in the Oregon country, was built northeast of Vancouver, probably on the edge of what is now known as Mill Plain. It consisted of buhr stones made from native stone and the top stone, or runner had attached to it a long sweep to which were hitched ponies for power.

"From the best obtainable information it is learned that wheat was next ground at Fort Colville, northwest of Spokane, apparently in the year 1833. If water

power was not immediately used, it certainly was soon after that year.

"Between 1832 and 1835 Hudson's Bay Company built a water power mill on a small stream where it empties into the Columbia about five miles above Vancouver. This mill had French Buhrs of best designs, shipped around the Horn. Water wheel and other parts were constructed on the ground from rough timbers.

"Very soon after the introduction of wheat at Vancouver, Hudson's Bay Company employees were encouraged to grow wheat on larger scales on what is now French Prairie, south and west of Oregon City and contiguous to the historic community known as Champoeg.

"The first mill to be built in what is now Oregon made its appearance at Champoeg in 1835, built by Webley Hauxhurst, and a grist mill followed at Salem in 1840; another in Polk county in 1845; one at McMinnville in 1853. Southern Oregon saw its first mill at Ashland in 1855.

"Southeastern Washington's first mill was built

26

HARNESS SHOP. The wheat farmer seldom took a trip to town without a visit to the harness shop. There was always something to repair or new equipment to buy. H. J. Eggleston's store was decorated as if for the 4th of July. But the date was actually November 21, 1911.

Courtesy, The George Lawrence Co.

The A. C. RUBY CO.

BREEDERS, IMPORTERS AND EXPORTERS OF

Percheron, Belgian, English Shire and German Coach

Stallions and Mares

Just home from Europe with a special train load of the best stock that money would buy in Europe. This shipment, with our stock on the Base Line Farm, will give our customers over 150 head in all to select from. We have a fine line of Percheron, Belgian and Shire mares in the lot, many of them gold medal winners in their native country. Will make close prices at the barn. Watch for our exhibit of 60 head at the leading fairs, and if you are in the market for one, make your wants known. Our prices are below competition, with a 4 year guarantee. Is this not more satisfactory than buying one without a guarantee? Your inspection or correspondence is solicited.

29th Street and Rose City Avenue, Portland, Oregon

NEGLECTED

Remember that the natural oil which leather contains is constantly drying out, and as it dries out the moisture creeps in.

Moisture dries and cracks it. Keep the moisture out by renewing the natural oil in the leather with Eureka Harness Oil.

When you keep on renewing the natural oil in leather with Eureka Harness Oil, the leather keeps soft and resists all kinds of moisture.

EUREKA HARNESS OIL

Practically keeps leather new. Leather treated with Eureka keeps its attractive appearance. Keeps its strength. Keeps its softness.

Eureka is little trouble to use. It doesn't soil the hands or the horse. It doesn't become rancid. It is the best oil for black leather. In fact, it has been selected with such care that it is now considered the *only* oil for black leather harness.

Eureka Harness Oil is equally valuable on black leather carriage or automobile tops; in fact, for black leather of any kind.

Sold by dealers everywhere.

Standard Oil Company
(Incorporated)

ADS FOR HORSEMEN. Advertisements in early-day Northwest farm magazines ran heavy to those promoting products related to the work horse. The A. C. Ruby Co. was one of the prime sources of draft horse breeding stock for the Northwest.

Kirby Brumfield

spread in the orchard. On young apple trees two forms are known, the ordinary crown gall and the hairy root. The latter occurs on seedlings, grafted story of how the best in seeds is the cheapest. **1911 Catalog Sent Free on Request**

P. O. ....
My dealer's name ....
is ....

"Gall Cure" Collars have saved the shoulders of many a horse and the dollars of many a horse ow take any chances. Get the "Gall Cure" Collar from your dealer.

W. H. McMONIES & CO.

24-26 UNION AVENUE   PORTLAND

READY TO WORK. Men and horses were lined up prior to a new day of harvesting. Feeding, grooming and harnessing the work animals was the first chore of the day.   *Courtesy, The George Lawrence Co.*

by Marcus Whitman at Waiilatpu, near Walla Walla, in 1840. This was a primitive water power plant.

"By 1871-73, mills operating in the so-called Oregon country numbered 23.

"The first American colony on Puget Sound erected a crude grist mill at the falls of the Deschutes in the village of Tumwater, near Olympia, in 1846. Later a better mill was erected there to be known as Ward & Hayes' Mill.

"Flour milling in Seattle proper dates from 1864 when Yesler installed a small set of buhrs and ma-

chinery near his sawmill to grind grain for the scattered settlers. Prior to that time Seattle was importing terrifically costly flour from Chile."

One of the early centers of wheat growing activity east of the mountains was Walla Walla. Dr. Marcus Whitman got things started when he grew wheat as early as 1837 at Waiilatpu. Following the Whitman massacre in 1847, the Walla Walla Valley was practically cleared of white men until after the Indian Wars of 1855-56 were settled. In 1857, Fort Walla Walla was established with Col. Steptoe in charge.

MITCHELL WAGON. The Mitchell people used this detailed ad to point out why their wagon axles and running gear were the best.

Courtesy, Kirby Brumfield

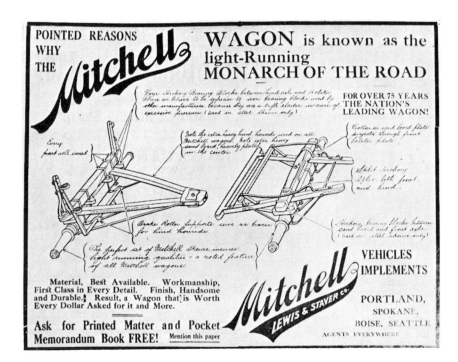

SHIRE STALLION. An imported, seven-year-old Shire named Rothwell Forester is shown at his fancied-up best. As if this wasn't enough the layout artist for Western Farmer magazine used his paint brush to eliminate the background and highlight the fine points of the animal before publishing the picture. This large breed came from England and resulted from Henry VIII's command that all horses less than five feet tall must be destroyed as useless.

Courtesy, Oregon State University Archives

HE'S THE BEST. Ike Horton of the La Crosse, Washington area almost bent over backwards with pride when showing off his ranch stallion. Farmers used to spend hours discussing the relative merits of their horses—their pulling power, their intelligence and their endurance.

Courtesy, Charles Freeburg

By 1864, the Walla Walla Statesman sparkled with pride when it was noted the local flour was superior to that produced in the Willamette Valley. The newspaper article also mentioned there were five flour mills in operation in Walla Walla.

H. M. Travis recorded in the August 26, '56 issue of the Spokesman Review that, "The first crop in the Touchet Valley was raised by Israel Davis on Whiskey Creek. It was harvested by cradling and threshed by trampling with horses. On a windy night, Davis's hired man cleaned 1000 bushels by pouring the grain from pails from a scaffold to the ground. The breeze blew away the hulls.

"It had been intended to use this precious wheat for seed in the spring, but because of the long hard winter of 1861-62 it had to be sold for food for people as well as livestock. Lots of the grain was ground in home grinders and used as cereal and coffee. People were glad to buy it for $2-2.50 a bushel, and miners paid $1 a pound for flour."

Farming expanded so rapidly in the Walla Walla district that by 1867 there was a surplus of wheat. The building of the Baker railroad in 1871-74 eased the surplus problem. The wheat was shuttled over the rails and down flumes to the Snake River port at Wallula for shipment to Portland.

Wheat growing in Oregon's Umatilla county is reputed to have had it's beginning in the early sixties when two sheepmen, a Mr. Thompson and Jerry Barnhart were bringing their flock back from summer range in the Blue Mountains. They discovered an excellent stand of grain had grown and matured in their sheep corrals. The plants had sprouted from kernels which spilled onto the ground from feed bunks earlier in the year.

A booklet honoring Umatilla counties first cen-

PLENTY OF HORSEFLESH. The grandstand of this 1922 Western Washington fair was jammed when these draft animals competed for the judge's favor. The judge carried in his mind a visual image of the ideal draft animal. It was his job to find the horse in the class that most nearly compared to the ideal.

Courtesy, F. L. Ballard, O.S.U.

HORSES COMPETE FOR BEAUTY PRIZE. The main street of Moro, Oregon became a show ring during the stock fair held October 10, 11, and 12 in 1910. The handlers were slickered up almost as much as their steeds were. The beef animal at the entrance of the livery stable must have felt rather out-of-place among all those horses.

Courtesy, Sherman County Historical Society

PROUD POSE. The 18th Annual Spokane Interstate Fair running from October 2 to 8 in 1911 featured some of the finest draft horses in the country.

Courtesy, Oregon State University Archives

BEAUTY PARLOR. Fitting and showing horses was an art. It was regarded important enough that most agricultural colleges featured courses on the subject. It was almost like going to a beauty school. The students in this Washington State University class were giving their steeds the full treatment. When the job was done the horses had been bathed, brushed and braided. Hooves were cleaned, noses wiped and ears checked for traces of dirt. The horse on the left sported the latest in hair coiffures.

Courtesy, Washington State University Archives

THEREBY HANGS A TAIL. The horse's tail also came in for its share of attention. A draft horse never entered the show ring with a natural flowing tail. Instead it was braided and tied and bedecked with ribbons. The handlers shortened the tail in this manner for two reasons. First, it gave the horse a blockier appearance. Second, it revealed the legs so the judge could see whether they were straight or not.

Courtesy, Washington State University Archives

BUSY PLACE. The George Lawrence Co. of Portland, Oregon was a very busy leather factory during the heyday of the work horse. The company no longer produces harnesses nor saddles. Their prime leather goods now are aimed at the sportsman with a big volume in holsters. In more recent years a large share of Lawrence business has shifted into automotive parts.
Courtesy, The George Lawrence Co.

ABOVE: A BIG DISKING OPERATION. Men were certainly outnumbered by mules in this scene. Fifty mules are harnessed to disks and six men did the driving. With the exception of an occasional clod, the ground had been worked to a mellow texture ready for drilling.
Courtesy, F. L. Ballard, O.S.U.

tury of progress was published by the East Oregonian in 1962. It reported that, "By 1868 several settlers in the vicinity of Weston had small fields of wheat. In the same year the first threshing was done with a small horsepower machine by William Courtney.

"William Switzler in 1876 plowed up the first of the bunchgrass land north of Pendleton and seeded it to wheat. After 1878, the hills in this region were rapidly converted into farms. The rush for land resembling in some respects the earlier rush to the gold diggings. With the coming of the Baker railroad up the Walla Walla Valley and completion of the Hunt railroad to Athena and Pendleton in the middle 80's the main wheat belt of the county came into its own."

Proud of their Umatilla grain product a local newspaper said, "The quality of wheat raised in Umatilla county does much to add to the reputation of the flour, which indeed, has a world wide reputation for strength and color as is evidenced by the fact that at the Chicago Exposition the medal was awarded the flour manufactured by the Pendleton roller mill against competition from the entire world."

Few of these early mills remain today. As transportation improved the need of a mill in every community diminished. It became more economical to concentrate milling activities in larger plants. These early day mills which were powered by steam, water or animals added almost as much romance to the wheat industry as the production of the wheat itself.

ACQUIRING LAND

Flour mills, of course, could develop only as farmers grew the wheat and wheat could be grown only as the pioneer sod-busters secured land. There were several ways these 19th century families could acquire property. Most of the newcomers moved onto 160 acre homesteads which the government gave them after certain conditions were met. The Homestead Act passed by congress in May, 1862 promised that any person over 21, who was the head of a family, and either a citizen or an alien with the intention of becoming a citizen, could become the owner of 160 acres of public land if he lived on the property for five years and improved it.

It was possible in many areas to get an additional 160 acres as a "timber culture." The person acquiring the land had to plant 10 acres of timber. Naturally, few of the trees survived in the semi-arid country. In Adams county, Washington, one greenhorn actually planted his young trees upside down. His argument of "Well, I planted em, didn't I?" wasn't terribly convincing to the folks in the land office at Colfax.

Still another means of getting land was "preemption" or paying the government $2.50 an acre as a substitute for the five year residence requirement.

There was also a fourth method of acquiring land. That was to buy railroad property. The government had given railroad companies title to every other section of land for 40 miles on either side of the new tracks as an inducement to build westward. The rail-

JOHN DEERE PLOW. The 1830's was an important decade for the development of revolutionary farm equipment. Besides the reaper, the first steel plow was introduced. It was in 1837 that John Deere cut a combined share and moldboard from a mill saw and beat it into shape over a form hewn from a log. It greatly improved the quality of plowing. Most important was the fact that the soil didn't stick to the plow making the instrument easier for the animals to pull. Later, in the century, gang plows were developed which made it possible to plow the large wheat acreages of the Northwest.

Courtesy, John Deere Co.

roads in turn sold much of this land to the hopeful new farmers. Early prices ran from $1.25 to $5 an acre.

This desolate land, some of it free and the rest inexpensive, was drawing settlers from across the United States and around the world. Advertising and word of mouth cued millions of people in on the fact

SUMMER FALLOW. Growing a crop that requires more moisture than comes from the annual rainfall is a trick that inter-mountain farmers have been performing for many years. It is possible through a practice called summer-fallowing. In brief, summer fallowing results in half a farm being planted to wheat while the other half is not cropped. The latter lies fallow which is another way of saying it is resting and storing up moisture. The following year the story will be just the reverse. The land which had a crop will be summer-fallowed and the acres which had been summer-fallowed will be growing wheat. The process alternates each year in this manner. It boils down to each half of the farm producing a crop of wheat once every two years.

The plowing in the picture was done in the spring. The series of gang plows were turning under wheat stubble left from the harvest of the previous fall. During the summer the farmer frequently ran a weeder over the ground to kill any moisture-robbing weeds that may have started growing. In the fall the field was worked and drilled to a new wheat crop.

Courtesy, F. L. Ballard, O.S.U.

there was land to be had in the Western United States. Thousands came.

To most the country was bleak and uninviting, but they stayed. The October 16, 1967 issue of the Spokesman-Review from Spokane told of an old-world settler searching for a new life and possibly even for a pot of gold at the end of the rainbow. He found both.

"For one 1889 immigrant, Etienne Geib, there was not time to waste on any regrets and disillusionments over this godforsaken Wilbur sector, which was in direct contrast to his homeland, the tiny duchy of Luxembourg.

"Geib had landed at Ellis Island, headed west and, for a spell, worked in Chicago. When he landed at Wilbur, Washington, he had $16.90 in his pockets. Eventually he was to possess 6,500 acres of fine farm land and to be rated as a millionaire.

"When he arrived, Geib knew no English but some of his countrymen, who had preceded him, befriended him. Geib later bought land from the Central Washington Railroad, which owned many sections of land along its right of way. He built himself a 12 by 12 foot shack and subsisted on trapped jackrabbits, potatoes and bread.

"Recently, Geibs' son Rudolph, a resident of Wilbur, recalled that first home with its dirt floor and rag rug, its stovepipe chimney and its homemade furniture."

Reflecting on the tide of immigrants flowing in, an article in the Palouse News of November 3, 1877 reported, "The immigration to the Palouse country has been so unusually large this season, we have about four consumers to one producer. The three saw mills in the county cannot supply the demand for lumber. There are now more than one hundred families with-

out shelter for the winter. Every woodshed or shelter of any kind is occupied."

Luckily, the Palouse winter of 1877-78 was a mild one. Everyone seemed to make it through the dreary months with none of their enthusiasm tarnished. The immigrants continued to stream in and the land offices did such a "land office business" that we had a new expression added to our language.

LIFE WAS TOUGH

Becoming a landholder was no guarantee of any easy life ahead. That came as no shock to most of the settlers. These hardy souls weren't expecting a soft touch. They were conditioned to tedious work and knew life would likely be a struggle. It was just that for many.

"Ten bushels to the acre was considered a good yield in the old days—and we were mighty thankful to

get any wheat at all." That's how 70-year old George C. Harter replied when asked about the "good old days" in an interview by the Ritzville Journal-Times newspaper in 1949. Harter had come to Adams County, Washington with his parents in 1888. They arrived in a covered wagon, built a sod hut in the sand hills area and struggled to exist.

Strangely enough, Harter claimed the greatest hazard the make-it-or-bust farmers of that area had to face was squirrels—hordes of squirrels. They were a constant menace to the early wheat crops and were the chief reason for many fields being so low in production.

"We'd plant 10 acres to wheat," Harter recalled. "When the crop was two or three inches high hundreds of those squirrels would take off the seed and ruin the field. If the wheat got to the heading stage the squirrels would straddle the stalk so the heads would fall to the ground where they could get at them. We'd find squirrel nests with the ground around them covered by wheat heads—as though they'd been thrown there by the handful," he remembered.

Jackrabbits were the culprits playing havoc with the early wheat crops in other areas.

Harter continued the interview by saying, "If it wasn't the squirrels, it was the heat. Month after month would go by without a drop of rain. Then we'd have a particularly scorching day—a hundred degrees at sun-up and getting hotter. Later that day a hot

breeze would drift in from the northeast. Before the farmer's eyes, his wheat, fresh green at daybreak, would fade to purple and then to brown. The scorched crop would look like a field of new-mown hay," he said as if it had happened yesterday.

RAINED OUT

Over in the Palouse country, there was a period when they weren't complaining about heat but too much rain. The darkest chapter of wheat history in that region occurred in 1893-94-95. There was so much rain in the fall of 1893 the grain couldn't be harvested. The little that was combined and sacked was lost in the fields. It was so wet there was no way to get it to warehousing. With no money coming in many farmers were unable to pay the interest or the principal on their loans. As debtors they were at the mercy of the creditors.

Prosser resident, H. M. Travis, recalled for the August 26, 1956 issue of the Spokesman Review that the wet weather resulted in much spoiled grain which led to a dubious practice called "stovepiping." He explained, "This was a process by which spoiled grain was mixed with good grain by placing a joint of stovepipe upright in an empty sack and filling it with the decomposed wheat. The intervening space was then filled with a good-and-later product and the pipe pulled out. The wheat buyers got wise to the practice and brought long metal tubes, which they called

NEW PLOWS. In 1909 Charles Kincaid was farming 1200 acres of land at Union Flats near Pullman, Washington. He rented the land from Jim S. Klemgard. Kincaid was riding the plow at the left of the picture. These plows were something new at that time. They featured wider bottoms. Kincaid traded in his three-bottom, 12 inch plows of previous years for these new, two-bottom, 16 inch plows. The stubble had trouble passing through the narrower plows so John Deere implement dealer Frank Campbell, standing in the field, sold Kincaid on the idea of trading in the old earth-turners for the new, wider style. They were known as Syracuse plows. Campbell must have been quite confident the new plows would be a success since he brought a photographer to the field with him. Atop the center plow was Rube Hollnebeck and on the right, Frank Murray, Kincaid's cousin. Surprisingly, the plows had no seats. The drivers handling the eight horse teams stood up all day.

Courtesy, C. H. Kincaid

TURNING THE STUBBLE. A 12-horse plow-team moves right along during this earth-turning scene in May of 1930. The location was 15 miles south of Wilbur in Lincoln county, Washington. The seat on the nearest plow was obviously never sat in unless the driver was interested in seeing where he had been. It was purposely turned around so the driver could stand on it easier.

Courtesy, Alexander Joss

'wheat punches' and pushed these down through the center of the filled sacks to detect if the grain was 'stovepiped'."

In 1894 the crop was fair but the price was low. The O.R.&N. Railroad Company was having a rough time of it too. It found itself in the hands of a receiver. The railroad was unable to grant reduced freight rates to farmers as it had done the year before.

The crop in 1895 was better but the price was still low. It was the year 1897 that put a smile back on the face of farmers in the Palouse wheat hills. Prosperity returned. Two things happened. Both price and crop yield were up. The price went up because of crop failures in other countries. The abundant crop made the year doubly sweet.

The only complaints came from the sack-sewers who had to work at a furious pace to keep up with the golden stream. There weren't enough threshers to handle the grain and harvest hands were scarce. Warehouses were inadequate, and temporary shelter had to be built. These were problems the wheat growers found to their liking. Many farmers paid off their entire mortgage with this one crop.

WHEAT IS KING

The business of wheat farming had its rough spots but, for those who managed to hang on the good years made up for the bad. On March 4, 1899, the Walla Walla Daily Statesman featured an article titled "Wheat Is King." Again, full of Texas-like pride the story started off in high gear with statements that would do credit to any present day Chamber of Commerce. Later in the story, facts were given which showed how wheat could be the pathway from rags to riches.

"As an agricultural district the Walla Walla Valley has just ground for laying claim to advantages of the greatest moment, and in many respects is unapproached by any other locality in the United States. While other agricultural productions in the county are of increasing importance, it nevertheless remains that wheat is the grand staple.

"The Walla Walla wheat district is the oldest of the three great divisions of the state and the product from the Palouse and Big Bend localities (the other two great wheat sections) is usually referred to in foreign markets as Walla Walla grain.

SUMMER FALLOW WEED-
ER. To keep weeds from springing up in the summer fallow
ground, an implement called a
rod weeder was developed. In
its earliest stages it consisted
simply of a bar sitting beneath
runners. The rod skimmed about
three inches under the soil surface cutting weeds off at the
roots. The first devices such as
the one pictured had a major
drawback. Every few feet they
became plugged with plants. The
driver then had to stop the horses
and clear the weeds off the rod.

*Courtesy, Sherman County
Historical Society*

TWO BARS BETTER THAN
ONE. A later development included two bars about three feet
apart. They both sliced under
the soil but not at the same time.
By walking forward on the
plank, driver Dick Morgan could
tilt the weeder down, sinking the
front blade into the earth as in
this photo. When this cutter bar
became fouled with weeds, the
driver merely stepped back to
the end of the board shown
sticking up in the air. His weight
threw the front blade out of the
soil and dropped the rear blade
in. With the front end elevated
the weeds and debris soon shook
off. The driver tilted the weeder
back and forth as he made his
rounds.

The final answer came when
a rotary rod weeder was developed. The rod on this implement
revolved underneath the soil.
The rotary action kept the weeds
from hanging on. The same tool
with improvements is used in
present day summer-fallow operations.

*Courtesy, Sherman County
Historical Society*

RIDING HARROW. Harrowing usually involved a lot of walking. The driver at the right was following the normal
procedure of hoofing it along with the horses. The man at the left decided there must be an easier way. He found that
riding had much more to be said for it than walking. He drove his team from the back of a riding horse.

Courtesy, Sherman County Historical Society

WEED KILLER. The name given to the two-bladed rod weeder was "slicker rod." It's easy to see how the rear rod is digging its way a few inches under the surface of the soil as the team of six pulls the implement around the field. Also worth noting is the hookup of the horses. Specialists in leverage worked out the eveners behind the animals so each four-legged worker pulled an equal load.

Courtesy, Washington State University Archives

"The total wheat crop of the state of Washington for 1898 was in the close vicinity of 20,000,000 bushels, of which nearly one-quarter was raised in the Walla Walla Valley. Harvesting commences in the Walla Walla district early in July and continues northward in the state, often until winter sets in November.

"Machinery of the latest development is employed in handling the big wheat crop in the district. Both the header and the binder are in use and a number of mammoth combination machines are operated, which cut, thresh and sack the grain at once.

"A majority of the wheat in the district is fall sown. Summer fallowing is generally and usually nec-

essary at times to rid the land of foul growth though it has the objectionable feature that under increased cultivation the rich soil produces too-heavy growth which lodges of its own weight.

"The wheat crop of 1898 approximated about 4,000,000 bushels as compared with 3,500,000 bushels in 1897. About 1,000,000 bushels of the crop of last year remains unsold, the big farmers holding the grain in anticipation of higher prices. Most of this grain is in sacks either stored in warehouses or in covered sheds at points along the railway lines located in the valley.

"The great majority of the farmers in the Walla

MULTIPURPOSE RIG. One way to speed up seeding was to use three drills instead of one. Hooking them all into one outfit also saved labor. This home-made rig looked like it was spread out over half the county but it was extremely effective at getting a great deal of seed into the ground in a hurry. The first string of harrows running under the wagon frame broke up the crusted summer-fallow soil. The three drills put the seed into the ground, followed by another line of harrows packing the earth around the seed so it would sprout faster. The outfit was used by Dwight Misner on his farm near Ione, Oregon in 1924.

Courtesy, F. L. Ballard, O.S.U.

DRILL TEAMS. With Moro, Oregon as a backdrop, the Lou Peetz outfit disks, drills and rolls in one sweep of the field. Drilling was done in the fall. The field had been in fallow all summer. It was at rest and hadn't grown a thing so it could save up moisture. The exposed wheat stubble on the surface of the ground was left that way purposely. It helped prevent wind erosion and the practice was called trashy fallow. The summer-fallow program was and still is practiced in low-rainfall, wheat-growing areas.

Courtesy, Sherman County Historical Society

Walla Valley are now in a better financial condition than those of most any other locality in the great northwest. This consequent upon the fact that the bountiful harvest which they have enjoyed during the past few years have been combined with big prices for grain. Three years ago many of them were heavily involved in debt, not only for the land which they held, but for the necessary machinery and horses with which their crop was harvested.

"As one of the many instances where Walla Walla farmers have prospered, the case of Charles Pickard may be referred to. Mr. Pickard came to the district ten years ago and engaged himself to Mr. W. H. Babcock at a salary of $30 per month and board. Mr. Pickard worked for Mr. Babcock three years at the same wages and during that time succeeded in saving enough money to purchase a small tract of land. He gave up his position after three years work and started on a small scale, raising a crop of wheat on his own farm during the year following.

"Regardless of the low prices prevailing during the hard times which followed, Mr. Pickard continued to acquire more land, purchased more machinery and made needed improvements. All this necessarily involved him heavily in debt. Last year he harvested an average crop of 25 bushels to the acre from 2800 acres of land and with the proceeds paid off all of his indebtedness, including an amount of $28,800.00 which he owed to one man, who had provided him with needed funds during the years of depression preceding 1896. Mr. Pickard is now reputed to own property to the value of $40,000, all of which he has acquired by raising wheat on his lands in Walla Walla county. He has the largest barn on the Eureka Flat; owns a full complement of farm machinery, including 18 drills, headers, threshing machines, etc., and has acquired them all during the past ten years, obtaining his start by working as a farm hand at $30 per month.

"Among the farmers of the Walla Walla Valley, who harvested more than the average number of acres

NO SOFT JOB. None of the early-day equipment featured soft seats to sit on, least of all the grain drills. In fact, they didn't have seats at all. If they did they were disregarded. Most of the drivers developed their sense of the balance to a fine talent as they stood atop the drill. This drilling was done in 1910 near Ritzville, Washington.

Courtesy, A. M. Kendrick

SEEDING WHEAT. The first six-horse, nine-foot-wide drill in the Colfax, Washington community was photographed with Philip Buffington doing the driving. According to the inscription on the back of the drill its trade name was "Kentucky."

Courtesy, Mrs. Glen Epler

may be mentioned, Mr. W. H. Babcock, who had 5300 acres in wheat last year; W. Reser, 4500 acres; Charles Pickard, 2800 acres; C. B. Upton, 2200 acres; Ed Bradbury, 2200 acres; George Struthers, 2000 acres; A. H. Crocker, 2000 acres; O. N. Wheeler, 1500 acres; C. R. Wilson, 1400 acres, and T. Welch, 1200 acres.

"The general average of yield last year was 25 bushels per acre and the average cost of raising the wheat, while varying in different localities, was 25 cents per bushel.

"One of the large operators, who does not wish his name mentioned, gave the following as the record of cost which he kept last year, when he received an average crop of 25 bushels to the acre:

	Per Acre
Plowing	$1.00
Harrowing (twice)	.25
Seeding, Hauling and Vitrioling	.50
Seed Wheat	.50
Heading	1.25
Threshing	1.25
Sacks	.65
Hauling to Warehouse	.33
Total Cost Per Acre	$5.73

"These figures bring the cost of planting and harvesting the wheat below 20 cents per bushel, but there are other items of expense which are not included in the above schedule. The wear and tear on machinery and interest on the money invested in the machinery and land is usually figured in with the cost of raising the grain. The average assessed valuation of wheat land in the county is about $30 per acre or from $10 to $50, according to location.

"Some of the big farmers who are operating in the foothills cannot raise their wheat at a cost of much less than 30 cents per bushel and that figure would probably also apply to the small farmer, who is com-

pelled to hire threshing crews and engage other labor and machinery than which he has on his own premises.

"W. H. Babcock, who raised the largest crop last year of any wheat grower in the county, in 1897 sold 90,000 bushels of wheat in Tacoma for $1.02 per bushel. He now has in store nearly 100,000 bushels of last year's crop, which he is holding for better prices."

The story makes it fairly obvious that the practice of holding wheat for bigger money has a long history It's also impressive how many of these early wheat growers were able to amass such large land holdings in a comparatively short time. The writer whose name was not listed cast a sharp bookkeeper's eye on the economics of wheat farming. He left the definite impression that the unpredictable occupation had its rewards. What seems most unbelievable is that he was able to secure so much confidential information from these hard-bitten sons of the soil.

A special Adams county pioneer edition of the Ritzville Journal-Times was published September 15, 1949 which gave a thorough look at wheat history in that county. It is representative of the wheat growing industry east of the Cascades. It had this to say:

"At first the settlers attempted to plow and crop their land every year, as most of them had done back in the midwest. It was not until about 1889 that they began to realize they must summer fallow their land every other year to conserve enough moisture for a profitable crop. While some farmers were harvesting the same land every year as late as 1900, the practice died out quickly after that.

"Early-day farming in Adams County was hazardous proposition, with uncertain harvests and 40 to 50 cent wheat, until the first bumper crop in 1897 proved that wheat could be raised here successfully and profitably.

"About that time wheat-raising gained predominance over cattle ranching for the first time in the budding county. The big crop of '97 also was a major reason for the tremendous influx of new settlers at the turn of the century.

"Practically all of the early grain was sacked. Warehouses sprang up all over Adams County, but elevators still awaited the day of bulking. At the peak of harvest, several hundred wagons a day came rattling into Ritzville and other points to form lines two or three blocks long before the warehouses. In the early 1900's Ritzville billed itself as the 'greatest primary wheat-shipping point in the world,' claiming more than 2,000,000 bushels were dispatched after a good year.

"By 1912, however, additional railroads had come into the county, spawning more warehouses, and many farmers who formerly came to Ritzville found shorter hauls to closer warehouses. Annual shipments in Ritzville were reduced to about 1,000,000 bushels.

"In 1905 the farmers made their first organized bid for higher prices by agreeing to withhold their crop from the market until they were assured at least 65 cents a bushel.

"At the same time they were waging a constant struggle to have railroad shipping rates reduced by 60 per cent. It cost about 40 cents a bushel to raise wheat because of advancing costs of machines and labor, the farmers argued, and the average local market price over a 10-year period had been—40 cents a bushel. In 1908 harvest wages were; sack sewer $3 a day; box driver, $1.75; loaders and stackers, $2.25;

DRILLING HOME A POINT. Early manufacturers knew the value of advertising. Space bought by these companies in magazines such as the Northwest Agriculturist and other farm monthlies of Oregon, Washington and Idaho told the story of why their implements were the best.

Kirby Brumfield

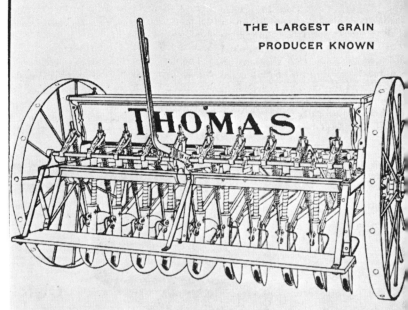
header driver, $2.75; picking up sacks, $3, and separator tender, $5. Unimproved land was selling for $8 to $15 an acre and improved land $14 to $35.

"The 1910 census gave Adams county a population of 10,920. The county had 16,000 horses and 1,700 mules and was farming with three-bottom gang plows, headers and steam-operated threshers and 32-horse ground-powered combines. Good land was up to $50 to $65 an acre.

"Weeds, no problem at first, appeared early in the 1900's in the form of Russian thistle and Jim Hill mustard. They became a serious menace between 1910 and 1915, robbing the soil of moisture and food and forcing farmers almost to pulverize their fallow by frequent plowings.

"Early-day weeders included a blade type, with the blades fastened at an angle to cut the roots underground. But they were spaced so closely as to plug up almost continuously. Later the stationary rod

41

WASCO MILLING CO. Much of the flour from this mill was exported to Japan and China. It was built in 1900 for $75,000. W. M. Barnett held the controlling interest in the mill which cost him $70,000. *Courtesy, Gordon Hilderbrand*

FIRE TAKES ITS TOLL. This was all that was left of a Colfax, Washington flour mill after fire swept through the building near the turn of the century. This salvage crew was finding very little that could be saved. It's likely this mill was originally a barn, but later adapted to the job of grinding grain. With so much dust in the air and on the timbers, fire was always a threat in a milling structure such as this.

Courtesy, Bill Walters

weeder was introduced—a gas pipe bolted to standards fastened to a sled-type frame. But these had to be dumped about every 20 feet and dragged the ground up badly.

"The solution came with the revolving rod weeder, still used with improvements, which allowed the trash to turn off as it moved along. Morning glory and blue lettuce have appeared in the past 15 or 20 years, but the chemical 2,4-D appears to have demonstrated it can handle them.

"Hard Adams county wheat, in great demand by millers, won international honors at expositions and fairs. P. R. Clark of Ritzville, as one instance, captured first place with a sampling of bluestem at the International Dry Farming Congress in Alberta,

Canada, in 1912. Three years later Adams county swept the top three prizes in white spring wheat in the International Wheat Show in Wichita, Kansas.

"Increased prosperity swept Adams county in 1916 as the 1st World War forced wheat above one dollar a bushel. By the following year it had climbed past two dollars. As land values soared, Ben Grote, largest rancher in the Walla Walla country, paid Emmet Hubbs $160,000 for 2100 acres southwest of Ralston, and Benjamin Berry about the same amount for 2000 adjacent to the south.

"Only a handful of headers and threshers were still in use. Motor-driven combines were replacing ground-power rigs, which would almost disappear by about 1925. The first Holt 75 Caterpillar tractor came into the county. Huge and cumbersome, they involved hours of mechanical work and frequently lost their awkward tracks.

"One old-timer, Levi Sutton of Washtucna, recalls they offered the driver his choice of two speeds: 'Slow and damn slow.'

HELIX MILLING CO. With a milling capacity of 400 barrels a day this Helix, Oregon flour mill was farmer owned. The bulk storage capacity of the four concrete tanks was 60,000 bushels. The year was 1918.

Courtesy, Oregon State University Archives

SANDOW FLOUR MILLING CO. This was another flour mill at Wasco, Oregon. It was owned by Walla Walla, Washington interests. Its construction was started in 1899.

Courtesy, Gordon Hilderbrand

"The first self-propelled combines were introduced during this period. In 1915 the first three were purchased by Henry, John and Jake Schoesler, Jack Danekas, and Dan Scott and Ben Gillespie. Instead of a header wheel they used one header 'track' for propulsion.

"Another important change which began making itself felt between 1915-20 was from horses and wagons to trucks. Tied into this was the widespread appearance of graveled roads. The first trucks didn't have enough power to get into the fields and horses still had to haul the wheat to the farmyard or a road.

"In 1920 Adams county still had 15,939 horses and 2,239 mules. But trucks and tractors, both of which were greatly improved during the '20's, began inevitably to replace the picturesque teams. As the horses disappeared, so did hundreds of miles of Adams county fences.

"The census of 1920 revealed that farm tenancy had increased 58.5 per cent during the previous ten years. In 1910, 275 of the county's 1,263 farms were operated by tenants. In 1920 tenants operated 436 out of 1,084 farms. The reason though, was not adversity but prosperity. Crops had been good and prices high during the boom of World War I, and many

WHEAT CAPITAL. In 1908 Ritzville, Washington was known as the bread basket of the world. It was the world's largest primary receiver of wheat. The panoramic camera shows how the town was built around the flouring mill and warehouse area. A. M. (Bert) Kendrick reports that homesteaders brought their wheat in from as far away as 50 miles. Some, he said, spent two days getting to Ritzville. After unloading the grain most teamsters unhitched their horses, tied them to the grain-hauling racks and fed them. Some of the men inclined toward the bottle would then get totally drunk, stagger back to their outfit, somehow hitch them up, flop down in the back and let the horses guide themselves home. Kendrick said they always got there.

Courtesy, A. M. Kendrick

1ST IN ADAMS COUNTY. John Gillette had the first horse-pulled combined-harvester in Adams County. The year was 1898. Farmers came from miles around to see the machine in operation.

Courtesy, A. M. Kendrick

early-day farmers had cleared enough to retire on, leasing their places to tenants on a sharecrop basis.

"The boom following World War I drove wheat to an all-time record of $3 a bushel in January, 1920. But two years later, as the recession of the early '20s set in, wheat was back to about one dollar and Adams county, sweating out a hot, dry summer, had only 1,770,000 bushels to show for its labors instead of the three million-plus it had produced in each of the four previous years.

"The 'wobblies' were causing trouble too. Professional agitators for the Industrial Workers of the World, they constantly beset harvest crews to strike for higher wages. One local farmer complained: 'If the going wage for a driver is $6 a day, the I. W. W. sets a strike for $8. If the wage is $8, he agitates for $10.'

"In 1923, though, Adams county bounced back to produce a terrific six million bushel crop, the highest on record. That year it passed Walla Walla county in Washington wheat production to rank third behind

Whitman and Lincoln, a position it has held most of the time since.

"Farm mechanization continued through the '20's and so did two other major transitions—from spring to winter wheat and from sacks to bulk.

"Many Adams county farmers had always preferred winter wheat to spring, because it enabled them to get their planting out of the way and free their time for plowing, weeding and other spring chores. The introduction of Turkey Red, an ideal winter wheat for areas of small rainfall, and succession of good fall rains encouraged a widespread swing to autumn plantings.

"Sacks went out during the '20s, too, as farmers found it much more economical to truck their wheat in bulk form. The familiar elevators began springing above the horizon. Moldboard plows, which laid the stubble on its side instead of turning it completely under, came about the same time. They have become an important factor in fighting erosion.

"The deep-furrow grain drill, introduced about 1927, was especially useful in the southwestern part

DRILLING AND HARROWING. The two operations of drilling and harrowing always went together. Once the seed was in the ground the harrow firmed the soil around the wheat kernels assuring better and quicker germination.

Courtesy, Sherman County Historical Society

of the county to plant seeds deeper into the subsoil moisture.

"But ever since the turbulence of the early 1900's settled down, Adams county's population had been shrinking as its farms grew larger. The 1,263 farms in 1910 had declined to 1,084 in 1920 and to 818 by 1930—a decrease of 33 per cent. The number of horses had dropped from 15,900 in 1920 to 8,800 in 1930. Nearly every week the Journal-Times carried an advertisement headed, 'Having decided to switch to tractor farming, I will sell my horses at public auction ...'

"Adams county emerged from the depression almost wholly mechanized. Three men with a tractor, combine and truck could handle a harvest that required a crew of 20 men less than 30 years before."

THE WISH BOOK. Most early-day wheat farmers and their families lived a somewhat isolated life. Many of the farm and home supplies were ordered from the Montgomery Ward or Sears Roebuck catalogs. The children often called them "wish" books because they were full of so many things they wished they could have had.

Kirby Brumfield

WALLACES' FARMER   1913

Drawn expressly for
Montgomery Ward & Company
by Charles Dana Gibson

Charles Dana Gibson is America's greatest artist.

He lives in his art—for art's sake.

The lessons of his pictures have punctured more illusions, intensified more sympathies and levelled the loves of high and low to the common standard of humanity.

We asked him to present through his art the message of Montgomery Ward & Co.

He has drawn two great pictures; one is called "Friends for Forty Years," which will appear later in this publication. The other drawing, which appears on this page—"My Father Loved that Book."

Do you get his meaning?

Her father *loved* that book because it made possible more comforts for his family at less cost than he could secure them elsewhere.

Whether he needed foods or furniture, clothes or curtains, machinery or farm implements, he could buy them from Ward's with all the extra profits cut out.

Have you traded at Ward's? Are you one of the millions or more people who have learned the lesson of economy by studying Montgomery Ward & Co.'s Catalogue.

If not, write us today. Copy of the great book will be sent you absolutely free, and without any obligation on your part.

Fill out the coupon. Send it today.

MONTGOMERY WARD & CO.
Chicago   Kansas City   Ft. Worth, Texas

SIMPLY SIGN THIS AND SEND TODAY
Montgomery Ward & Co.
Dept. N18, Chicago.
I accept your offer to send me your 1,000 page Catalogue without cost to me.

Name ....
Address ....
Town ....
State.................. R. F. D...........

My Father Loved that Book"

PLAT OF
TOWNSHIP 16 NORTH, RANGE 43 EAST, W.M.
WHITMAN COUNTY, WASH.
Scale 2 Inches = 1 Mile.

References:
Railroad
Wagon Road
Creek
Trail
Numbers with circles (7) indicate the adjoining page or the number of page on which a larger scale map will be found.
Lakes O

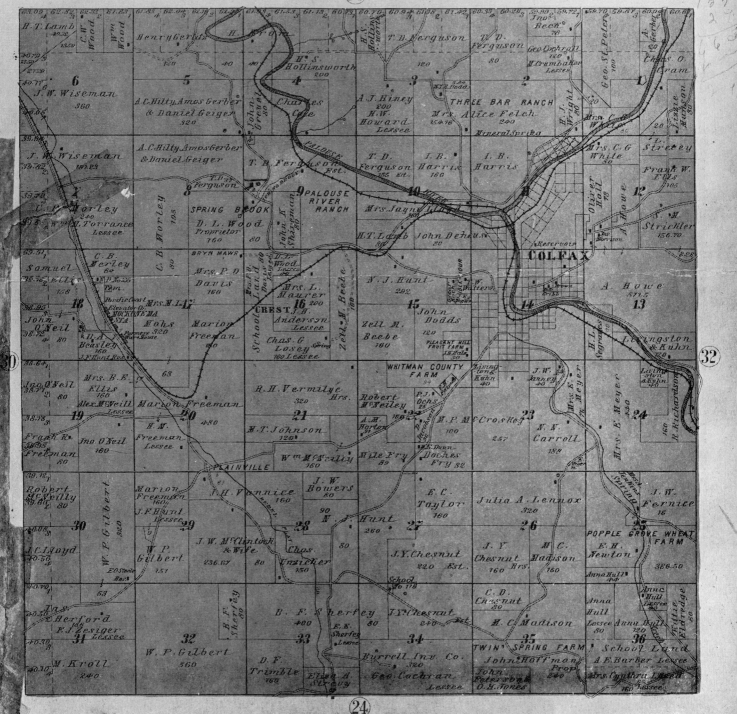

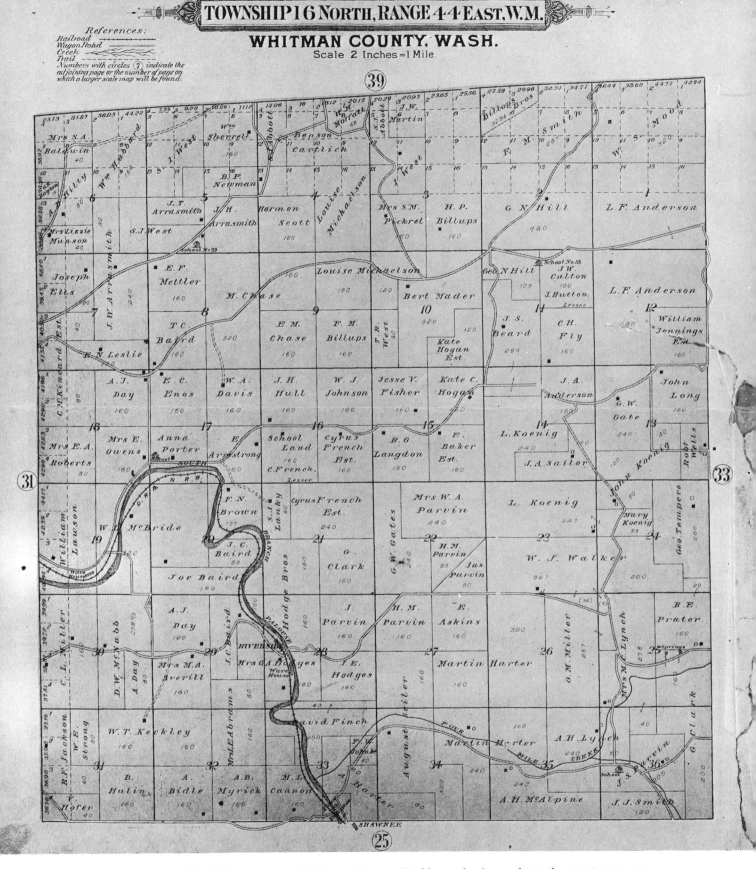

MAP OF WHEAT COUNTRY. A portion of Whitman County, Washington land map shows the country was completely settled by 1910. Whitman County took an early lead in Northwest wheat production and still claims that distinction. In fact, it is the highest wheat producing county in the United States.

Courtesy, Dee Doty

PROMOTIONAL CAMPAIGN. "From field to biscuits in 20 minutes" was the slogan behind this 1912 promotional campaign for Pure White flour in Waitsburg, Washington. The wheat was brought into town from the Zugar farm. There it was run through a miniature mill. Another piece of special apparatus for this demonstration was the tiny oven in the picture. The baking was done by several men from the bake shop uptown. The fellows gathered around the oven were mill hands.

Pure White flour was a product of the Preston-Schafer Milling Company. The promotional campaign became a Waitsburg community event. On this July day in 1912 townspeople and mill workers gathered to witness the hoopla. Dale Preston, president of the milling company, was on hand. He sat behind the driver's wheel of the first car. Driving the second car was Walter Woods, local farmer. Next to him in the Franklin car and looking very proper was Mr. Wheeler, editor of the Waitsburg Times newspaper. The engine in the background moved boxcars in and out of the mill siding as they were loaded with flour. The train operated between Pasco and Spokane, Washington.

Courtesy, Paul S. Hofer

HOUSE STYLES. These homes owned by both towns people and farmers in Sherman County, Oregon near the turn of the century were among the most elegant of the area.

Courtesy, Sherman County Historical Society

READY FOR HARVEST. There are millions of individual kernels of wheat in this field. Historically, one of man's greatest challenges has been to find a way to separate the kernels from the wheat plant. A sickle or scythe and a flail provided the only answer for centuries. Cyrus McCormick's reaper in 1831 opened the door to an avalanche of developments aimed at simplifying the process. By the late 1800's wheat had become a big item in the agriculture economy of the Northwest, especially because of new machines that made it possible to plant and harvest large acreages of the golden grain.

Courtesy, Washington State Historical Society

Chapter III

The Lore Of The Harvest...

THE FIRST WHEAT IN OREGON AND WASHINGTON was in small plantings, strictly for home use and cattle feed. There's no denying that such primitive harvesting equipment as the cradle-scythe was used in those earliest days. But that was when the Northwest was still thinly populated. It was before large droves of sod-busting pioneers flooded in to till the soil. It was before binders, headers, threshing machines, horses and steam made huge commercial wheat acreages feasible. The increasing size of farms and development of machinery to handle them were a parallel development in the Northwest.

Although the reaper of 1831 worked, McCormick saw much room for improvement. He kept tinkering and devising through 1835. Only then did he submit a patent application on his revolutionary invention. The experimentation continued into 1841. Then, in 1842 seven reapers were sold. Sales rose to 29 in 1843 and in 1844 the sales curve shot up to a total of 50 sold.

That was just the beginning. Moving his business to Chicago in 1844, McCormick and his staff kept seeking improvements. A rake arm was attached to the reaper in 1862 which raked the cut grain off the platform and to the side of the machine. This eliminated one man who had done the job. Up to this point the grain was still bundled and tied by hand.

A further step in the evolution of the reaper came in 1874 when a wire binder was developed. This machine tied bundles with bands of wire. Dubbed a "binder," the name of the machine described just what it did—bound the grain. Again it whittled away at the manpower needed by eliminating the hand binders who had ridden the Marsh-type harvester invented by C. W. and W. W. Marsh of Dekalb, Illinois. Seven years later twine replaced the wire in the binders, much to the farmers' liking. McCormick was by no means alone in the field. There were dozens of farm implement manufacturers in the country, each one adding improvements and new approaches to the harvesting of crops.

BINDING AND THRESHING

Binders were used extensively in the wheat harvest of the midwest, up into Montana and Idaho and to a limited extent in Washington and Oregon.

Even though they ranked far down the list in total popularity among Northwest harvesters, binders were preferred by some. They found strong support in the Spokane Valley of Washington and the Willamette Valley of Oregon for instance. They popped up here and there in both states.

Four horses provided the pulling power for the binder. These updated reapers cut widths of grain varying from six to eight feet depending on which model the farmer chose. If they cut much more than eight feet the side-draft became too much for the horses to easily handle. The eight-foot width

MARSH HARVESTER. A forward step in the evolution of harvesting machinery was the Marsh harvester in 1858, shown with its three-man crew. This machine employed continuous canvas aprons to raise the grain to a table. There, two men riding on the reaper bound the collected grain and threw the bundles over the side onto the ground.

Courtesy, International Harvester Co.

BINDING PALOUSE WHEAT. The driver always dumped the bundles of wheat in the same spot, forming a row of bound grain. This meant less walking for the men pitching the bundles onto wagons later.

Courtesy, Washington State University Archives

was generally the most popular. The height of the cut was regulated by the operator who had a seat on the back end of the machine.

As the grain was cut by the low-riding sickle bar a rotating paddle-reel gave the tall stalks a last minute nudge so they fell backwards instead of forwards. With this extra little push the stalks fell straight back, onto a revolving canvas belt. Its correct title was the platform drape. It carried the wheat at a 90 degree turn away from the sickle bar.

The tying mechanism was located almost four feet above this lower level. The platform canvas sped the wheat along its full length and shot it in between two revolving drapes set at a 45 degree upward angle. These were appropriately tagged the elevator drapes. By running the wheat between them in a sort of squeeze play the drapes left little doubt which way the grain was going—up.

The wheat went up and directly into the tying mechanism. When a certain amount of wheat was in place a needle was tripped. The amount was determined by weight and could be adjusted to larger or smaller size bundles. Most farmers went for the larger bundles to save twine.

Tripping the needle sent a long, curved, fingerlike piece of metal into action. It shot forward, curling around the stalks and carrying the twine with it. This put the twine in contact with an ingenious little device termed a knotter. It took the twine which now encircled the newly formed bundle and tied a knot. The needle quickly returned to its normal position and the bundle was dumped onto a bundle carrier. The carried held from four to six bundles.

An operator riding in the rear of the binder hit a foot lever to drop the gathered bundles to the ground. As the binder moved around the field the bundles were dropped at the same points on each round.

Most binder drivers weren't given much to deep thought. They had neither the time nor inclination. Handling the horses, adjusting the header height and keeping an eye on the machinery left the driver wishing he had three pairs of eyes and six arms. Certainly there was little thought of art or design drifting through his mind as he made his endless rounds. But in spite of himself he became something of an accidental artist as he dropped the bundles.

Looking down on the field from a higher elevation the bound grain formed eye-appealing patterns of flowing lines starting at the outer edges and converging into the center.

As the binder steadily clicked its way through the wheat a man called the shocker hitched up his pants, spit on his hands and went to work. Equipped with a pitchfork he speared the bundles into shocks. This was simply grouping five or six bundles into an upright position so they stood by themselves. One man could handle this job if the wheat wasn't too heavy, but more often it was a two man task.

The bundle wagons moved into the field next. This happened soon after the binding was done. Two men pitched the bound grain aboard the wagons. There were no sides to them, just a flat bed and an upright

TWINE BINDER. Cyrus Mc-Cormick's inventive mind didn't stop with the reaper. That was just the beginning of his contributions to the new farm machinery industry which he fathered. This was one of the first McCormick twine binders that began appearing on the scene in quantity during the 1880's. It was the fullest extension of the original reaper in that it cut and automatically bound the wheat. It required only one man to drive the horses and watch over the machine's operation. It cut on the average of 15 acres a day.

Courtesy, International Harvester Co.

THE BINDER BUNDLES. This was a typical binding scene on farms west of the Cascades. Many binders wer equipped with a carrier so they could hold bundles and drop them five or six at a time. The machine picture didn't have this advantage though and dropped each bundle as it was tied. It was a smaller binder and was pulle by three horses instead of four.

Courtesy, F. L. Ballard, O.S.U

OPENING A FIELD. There was no getting around the fact that opening a wheat field with a binder meant some of the wheat would be trampled down by both horses and machine on the first round. This photo shows that the first round had been cut with the second round in progress. The team pulled the binder in one direction on the opening round and in the opposite direction for the remaining rounds. The contrary direction of the binder on the opening round was necessary to cut the wheat in close to the fence on the left. The bundles of bound grain dropped on the ground during that first round had to be picked up before the second round could be cut. Some of the down grain is seen in the bottom right corner of the picture. A good operator could get the sickle bar of the binder under a large share of the trampled wheat so it wasn't all lost.

The wooden sticks between the three horses on the left served as dividers so the animals didn't crowd in too close to each other.

Courtesy, J. F. Abernathy Livestock Photo Co.

set of stakes at each end. A man on the wagon called a loader placed the bundles so none would tumble off on the ground as the wagon jolted its way to the threshing machine.

The next stop for the loaded wagon was right in beside the stationary threshing machine. More hand pitching followed as the bundles were fed into the separator. Oscillating knives at the throat of the machine cut the twine. If a threshing machine was not immediately available the bundles were brought in and stacked. Later, the separator was moved in near the stack and the threshing was done from the stack.

WATER BREAK. Binding and shocking were being done at the same time in this wheat field near Ontario, Oregon. The man in the foreground was carrying a bundle of wheat by hand. He may have carried each bundle to the shock in this manner although it would have been quite uncommon. The shocker normally used a pitchfork. The Ontario, Oregon area is known for its variety of crops. This was true when this picture was taken as wheat and corn were grown side by side.

Courtesy, F. L. Ballard, O.S.U.

BUNDLED WHEATSCAPE. Professional photographer R. R. Sallows caught the beauty of a field of shocked grain in this 1908 photograph. The wheatscape was near Goderich, Canada.

Courtesy, Oregon State University Archives

HEADING AND THRESHING

Throughout the bulk of Eastern Oregon and Washington in the late '90s and into the 20th century, heading and threshing was the standard harvest procedure. Heading was done by a header. The header didn't bind the grain. It merely cut off the heads along with eight to twelve inches of stalk and conveyed this into a wagon traveling alongside. The machine cut an average of 30 acres a day.

The header worked differently than most other farm machines. It was pushed instead of pulled. The business end of the machine moved into a field first.

BUNDLE WAGON. There was nothing fancy about the bundle wagon. It was simply a farm wagon which could be used interchangeably to haul hay or bound grain. Sometimes one, but more commonly two men pitched the bundles from the ground to the wagon. The driver helped stack the load so it wouldn't slide off on the way to the threshing machine.

Courtesy, Evelyn Terry

Providing the muscle were six horses which came along behind and pushed the contrivance. Sitting on the tail end, between the horses, was the driver or "header-puncher."

Compared to more conventional machines it looked somewhat strange moving down through a field. There was a little adjustment period for some of those early-day farmers as they got used to seeing the horses located behind the machine instead of in front of it.

The cutting swath of a header was usually 12 feet. The arrangement of the sickle bar, reel and platform drape was similar to the binder. There were even elevator drapes. Rather than piling the wheat into a space for binding though, it was carried straight out where it spilled into a wagon.

The header traveled on three wheels. Up to the front and left of the platform was the bull wheel. As the machine moved forward the bull wheel provided the power that ran the machinery.

To the right side and behind the platform was the grain wheel. It was the key support for the right side of the header. A solid steel pipe ran from the frame back to the pilot wheel. This third wheel not only helped support the header but was also used to steer it.

There were three horses on each side of the pipe. The left hand set of horses wasn't usually tied in as tightly as those on the right. Any change in direction was most frequently to the right. This made it easier for the horses to swing out to get the machine around those turns.

The header-puncher had a seat positioned over the rear pivot wheel. He half-sat and half-stood there with his feet braced solidly against the frame of the machine. Up between his knees ran a rod which was attached to the pivot shaft of the rear wheel. By moving his knees to the right or left the operator could guide the direction of the header. The skilled header-puncher had little trouble steering his machine so it took a full bite into the standing wheat.

WHEAT SHOCKS. Waiting for the trip to the threshing machine was this field of Flood wheat near Rosalia, Washington on August 22, 1915. Asahel Curtis periodically roamed over Washington and parts of Idaho pictorially recording such harvest scenes. His brother, Edward, was also a well-known photographer who specialized in filming western Indians under a grant from Theodore Roosevelt.

This wheat has been bound. Several men armed with pitchforks then moved through the field placing ten to fifteen bundles together in an upright position. This collection of bundles was called a shock. A shock did two things. It held the heads of wheat up off the ground so there was no chance of picking up unwanted moisture. Secondly, the shocks meant less stops for the bundle wagons as the pitchers loaded them.

Courtesy, Washington State Historical Society

While manipulating the header and horses so a full swath was cut, the driver also regulated the height of the cutter bar and platform by a lever in front of him. Much early wheat grew five and six feet tall. The resulting excess straw made it difficult for the thresher to do an efficient job of separating the wheat from the stalks. By reefing back on the lever and titlting the cutter bar up, the cut was higher leaving more of the stubble in the field.

The wagons which carried the headed wheat were called header-boxes. There were never less than three or four header-boxes per header in a harvest operation. The header kept them all busy.

The header-box had sides on it but they were "caddywampus." The left side of the box was almost five feet high while the right side was only one and a half feet high. The front and back end of the wagon assumed the appropriate slope which gave the whole rig a cockeyed look.

The dimensions of the larger header-boxes were nine by eighteen feet and eight by sixteen feet for the smaller ones. Usually one, sometimes two teams pulled a wagon.

Once a wagon was filled, the loader clambered onto the header spout waiting for the next header-box to move into position beneath him. Once it did he jumped down into the box.

The loader swung into action the minute both

THRESHING AT THRESHER. As the bundles were moved into the mouth of the threshing machinery by the conveyor, a series of sharp knives sliced up and down cutting the twine binding the wheat. This photo taken by Asahel Curtis on August 10, 1939 at Thresher, Idaho shows the grain being bulked. Hauling wagons were backed in under the bulk hopper, where they were loaded.

Courtesy, Washington State Historical Society

header and wagon began moving forward. It was his job to stack the grain so as to get the largest load possible. A good driver could make it easier for the loader by gradually moving his wagon forward during the loading process.

Holding the harvest equipment on the unbelievably steep Palouse hills of Washington was a tricky business. So difficult, in fact, that header-box manufacturers began expanding the width of the rear axles up to 12 feet. The added dimensions assured better footing on the precarious slopes. These expanded axles came to be known as "Palouse" axles.

Even then, on extra steep ground the wagons had a tendency to slip and slide. When this happened, it wasn't uncommon to chain the header-box to the header while loading.

Occasionally, a header-box driver got careless or his team got out of control and the wagon slammed into the header. For just such cases two wooden rollers were placed on the under side of the header spout to protect it. Many box drivers felt the sting of the header-puncher's tongue as he chewed them out for bumping into the header spout.

Most farm harvest machines start around a field from the outside and work in. Not so with the header. Its operation was just the opposite. It worked from

PITCHING BUNDLES INTO THE THRESHING MACHINE. Most Midwestern threshing scenes looked like this. The thresher was set up in the farmyard. The acreages of wheat were, of course, much smaller than in Eastern Oregon and Washington. Normally, the wheat fields of Midwestern farms were close enough that the bound wheat could be brought into the one-stop spot for separating. This was convenient for the diversified farming of that region since it put the straw pile next to the barn. Frequently the straw was blown directly into the hayloft of the barn.

Courtesy, J. I. Case Co.

TWO HEADERS DO JOB TWICE AS FAST. The picture shows how the header-puncher positioned his header rig so it loaded the wagons from front to back.

Courtesy, Sherman County Historical Society

the inside out. The header-puncher got the harvest going by taking a bead on the center of the field and cutting straight through to that point.

The header-puncher then started going around in circles. He worked his rig in ever-widening spheres from the inside to the outside. Among other advantages, this procedure eliminated sharp corners.

Opening a field meant a considerable amount of the standing wheat was trampled. It wasn't knocked down by the header itself, but by the team and wheels of the header-box traveling alongside. To save the wheat many farmers attached a huge canvas bag to the spout. It caught the grain until there was enough area opened up for the wagon to accompany the header.

This worked fine in the smaller fields. The bags weren't large enough for bigger acreages, so, the header-boxes had to be used from the start—in spite of the downed wheat.

There were some header-thresher farmers who used a binder to open their fields. If a binder was used to cut an opening path and center circle, the job was generally done a bit earlier in the year so the bound wheat could be used for hay.

Similar to a binding operation, headed wheat could be either stacked or pitched directly into a threshing machine. Some of the bigger operations headed and threshed at the same time. This necessarily involved more men and more equipment. It meant not only one but from two to three headers, seven to nine header-boxes and a threshing machine all on the premises at the same time.

If a farmer, on the other hand, was going to thresh from a stack of headed wheat he built his stack or stacks in the center of the field. They were set at the hub of the newly-opened circle. This reduced the distance the header-boxes had to travel. The wagons

could go from where they unloaded straight out to the header at the rim of the harvest circle.

Stacking called for the header-box driver to pull into the stack as close as possible. He and the spike pitcher then worked up a good sweat as they pitched the wheat from the header-box onto the mounting stack. On top of the pile a stacker placed the wheat as he "built the stack."

In very large fields, it was sometimes necessary to build as many as four stacks. If so, they were arranged in pairs so the separator could be set in the middle. This minimized the time of moving from one stack to the next. One setting of the threshing machine served for two stacks thanks to a wing feeder. When the first of the huge piles was finished the crew swung the feeder over to the adjoining stack.

Once the stacks were built and the threshing was about to begin a "derrick table" was put in place. It was set in convenient pitching distance to the feeder. Very simply it consisted of a large platform set on a heavy duty wagon. Four poles provided the support. One pole led down from each corner in tripod-fashion to a common center.

Over this platform, three tall poles were set, tepee style. They formed the derrick. A large pulley was fastened to the peak of the derrick and another was staked to the ground.

A rope or cable passed through these two pulleys with a team of horses hooked to the lower end. A Jackson fork to lift the wheat was on the upper end.

A forker jabbed the huge fork into the stack and gave a yell when it was set. That was the signal for the derrick driver to move his team forward. This moved the loaded fork over to the table where it was dumped. Expert forkers made sure the fork was placed the same way each time so the top of the stack stayed level.

Once the wheat hit the platform a pair of "hoe-down" men went to work. To rake the grain into the feeder, they used a tool that looked like a cross between a pitchfork and a garden rake. These men worked at top speed. It was a grueling job which explains why they worked for 15 minutes and rested for 15 minutes.

The mechanical feeder was a conveyor arrangement which moved the grain into the separator for threshing.

THRESHING

Headed or bundled, stacked or fed direct, all the wheat ultimately wound up in the threshing machine.

Many threshing machines traveled from one farm to the next during harvest time working on a custom basis. The youngsters on every farm looked at the arrival of the thresher and all the activity that followed as the great event of the year. These well-remembered queens of the wheat fiields sported such nostalgic names as Red River Special, Case, Advance Rumley, Reeves, and Pride of Washington to list a few. The latter machine was manufactured in Walla Walla, Washington by the Gilbert Hunt Company.

There's no doubt that these trusty threshers made is possible for wheat to mushroom into a multi-million dollar business in the northwest. The millions of bushels of the golden grain which passed through their innards served as a passport for many from serfdom to wheat-producing land barons.

No matter what romantic embellishments we attribute to the threshing machines though, the fact remains that working around them wasn't the most enviable job in the world.

Dust was a way of life with those old separators. Sack-sewers and hoe-down men, working at opposite ends of the machine, always tried to verbally cajole the wind into blowing away from them and taking with it all the dust and chaff that poured from the shaking and vibrating thresher. The breezes seldom cooperated, however, making life around a separator anything but comfortable.

Soon after work started each day, the workers' faces resembled chimney sweeps. If these dust-eaters could muster up enough moisture to spit it always came out black.

CREWMEN

Number one man on the threshing machine was the separator man. He was responsible for keeping the threshing machine running. He fussed over his charge like a mother cares for her baby. This master mechanic ranged around, under, in and on top of the thresher always on the lookout for trouble. Like a piano tuner's trained ear the separator man could know instantly if something was wrong just from the sound of things.

With an oil can in one hand he was continually dabbing lubricant on moving parts that seemed to need its cooling relief. There were always grease cups to tighten down and nuts that were loosening up. Frequently a separator man had to work late into the night making a repair or an adjustment so no time would be lost the next day. A lantern supplied his light.

Once the threshing machine knocked the grain kernels free the left-over stalks were called straw or chaff. This chaff flew out one end of the separator and the grain was funneled out the side. Principal characters in sacking the wheat were one or two sack-sewers and a sack-jig. They usually worked under an improvised shelter attached to the threshing machine to minimize dust.

An auger tube carried the newly threshed grain to the sacking area. The end of the tube split V-shape into two spouts. A sack was hung on the end of each spout. When one sack was filled, the sack-jig flipped a divider paddle which sent the wheat spewing into the adjacent burlap bag. He then took the filled sack off the hooks, gave it two or three jounces or jigs and set it between the knees of one of the sewers.

A sack-jig took a real chance if he didn't place the sack squarely in front of the sewer. The frequent

HEADER AND HORSES. This side view of a header almost looks like a cutaway drawing of the implement. Once the wheat was cut it fell back onto the revolving canvas drape which hurried the headed grain to the elevator spout. It was then up and out into the waiting wagon.

Courtesy, Sherman County Historical Society

DRIVING WITH THE KNEES. Driving a header was like driving no other horse-powered contrivance. The horses pushed, they didn't pull the machine. This made steering a bit more difficult. The problem was solved by setting the header on three wheels. The photo shows two wheels up front and a third in the rear. Facing the challenge of steering the outfit was the "header-puncher." It helped if he was shifty in the knees. Notice the rod extending out between the knees of the nearest driver in the scene below. The teamster, half-sitting and half-standing, steered the header by shifting his knees to the right or to the left. Strange as the system may sound, the skilled header-puncher could put his machine wherever he wanted it.

The rope slings are obvious in the two empty wagons. The headed wheat was piled on top of them. At the threshing machine the slings were pulled out, dumping the headed wheat onto a platform, where it was pitched into the feeder.

Courtesy, Sherman County Historical Society

penalty for not hitting the mark was a sharp jab in the leg from the sack-sewer's needle.

The sack-sewer sat on two full sacks of wheat. These fellows had a way with a sack needle. Their hands flew as they flipped a double half hitch around one end of the bag to form an ear on the sack. They next sent the needle back and forth across the top for nine to eleven quick stitches and finished up with two more half hitches to make the second ear. Not a motion was wasted. Then, with a deceivingly easy toss across his knees the sewer carried the 130 to 145 pound sack to a nearby pile.

The sacks were piled five high. There were often as many as a thousand in a setting and were a favorite play area for the kids—if the farmer permitted it.

An average sack-sewer could needle his way through about 650 sacks a day. Experts could run that figure up to 1000.

The first separators shuttled the straw out the tail end and dumped it on the ground. Unless carried away it piled up there and could eventually plug the machine. The straw-buck had the far from glamorous job of removing the straw. With a team and a fork he dragged the straw a short distance off, where it was piled.

REAR VIEW OF HEADER. The header-puncher was always positioned at the tail end of both horses and machinery.

Courtesy, Sherman County Historical Society

If the threshing machine was powered by a straw-burning steam engine, the straw-buck moved a portion of the straw in beside the tireless giant where a fireman steadily fed one forkful after another into the firebox.

The threshing machine manufacturers soon added conveyors and then blowers to their separators. These mechanically transferred the chaffy material out to the straw pile.

Also a part of the threshing crew was a "roust-about." He was responsible for just about everything no one else was doing. That ranged from trips to town for groceries and parts to bringing wood from the farmstead and splitting it for the cook.

SITTING DOWN ON THE JOB. Perched on the header spout like a crow on a fence, the man called the "loader" frequently assumed the position shown while waiting for an empty wagon to pull in beneath him. He would then jump down into the wagon box and the loading began.
Today's machinery seldom relies on belts or chains to supply power. Early machines used little else. Even the canvas drapes on the header spout were run by the twisted belt seen on the side.

Courtesy, Sherman County Historical Society

GROUP SHOT. Three headers and six header-boxes represent a fairly large unit. It was big enough that the wheat was probably being fed from the wagons directly into a threshing machine rather than being stacked first and then separated. The shape of the header-wagons or boxes was unique but functional. With one side lower, the header-box driver could move his wagon in under the header spout as the two conveyances moved through the field. The headed wheat cascaded from the spout into the wagon where it was leveled by a man equipped with a pitchfork.

Courtesy, Oregon State Historical Society

PICTURE POST CARD. Harvest scenes were frequently used on post cards. This picture card was sent to Jack Williams by his father in 1928. The father was working in the harvest pictured at Davenport, Washington.

Courtesy, Jack C. Williams

HEADER AD. One of the main suppliers of farm equipment for Pacific Northwest wheat farmers was Mitchell, Lewis and Staver Co. They had offices in Spokane and Portland. One of their ads in a 1920 Oregon Farmer magazine gave a pitch for the Avery header which they handled.

Kirby Brumfield

HEADING AND STACKING. There were no derricks to ease the unloading job on this farm. The headed wheat was unloaded by hand. Later, the threshing machine would pull in alongside the stacks and separate the grain. This approach to the harvest was less efficient since the grain was handled more often. Grain was lost as a result but it was the common practice for smaller outfis—ones that didn't have enough headers, horses or men to keep a threshing machine running full time as the wheat was brought in from the field.

Courtesy, Wayne Doty

UNLOADING THE HEADER-BOX. Action shots were seldom taken in the early days of farming simply because photography was also in its early stages. Films, shutters and lenses were not usually able to stop action but they certainly did in this photograph. This rare picture shows exactly how the header-box was unloaded with the help of the rope sling. As the derrick team at the left moved forward the headed wheat was dumped over onto the derrick table. The derrick table pictured was merely a wagon. Frequently it was a specially built platform set on top of a wagon. The men pitching straw on the stack in the background look a little hazy, as they were completely enveloped in a continual cloud of dust.

Courtesy, Sherman County Historical Society

MEN, MACHINERY AND HORSES. Scattered about on headers, wagons, engines, a thresher and even high atop an unloading derrick were the workers that comprised the crew of this header-thresher operation. They totaled 28. Today, two or three men with their modern self-propelled combines can do the harvest job which once required all these men and horses.

Courtesy, Sherman County Historical Society

KITCHEN ON WHEELS. The cookhouse in this setup was smaller than most, but no matter what the size it was a welcome sight to a crew of hard-working field hands morning, noon and night. There was never anything very fancy about these harvest cookhouses. They were slapped together as simply as possible and seldom sported a coat of paint. The cooks loaded the tables down with meat, potatoes, gravy, pies and a host of goodies that seem impossible considering the simple tools they had to work with. The ladies usually slept in a tent next to the cook wagon. It was 1906 near LaCrosse, Washington that this pleasant looking lady took a few minutes off from her cooking chores to pose for the cameraman.

Courtesy, Charles Freeburg

THE WOMEN IN HARVEST

When the engine whistle blew at mealtime, the most appreciated workers on the ranch were the cooks. Heaps of meat, potatoes, gravy, home-made bread, vegetables and desserts disappeared before the ravenous crews of 15 to 25 men.

The cookhouse was set on wheels. It was pulled from field to field as the harvest progressed. The mobile kitchen featured a wood stove at one end. Tables and benches lined both sides. Between meals dishes and supplies were kept under the tables.

Ventilation in the cookhouse was limited. What there was came from a couple of boards along the outside which swung up. With the sun beating down and heat generated inside by the stove, the cookhouse could more aptly have been described as a hothouse. The door at each end of the structure wasn't much aid in holding the temperature down.

There were usually two ladies handling dinner chores—sometimes a lady and one or two youthful assistants. They were hardy souls and put in as many hours a day as the men, and frequently more.

Sleeping accommodations weren't the greatest. At least there was a choice. It was either the floor of the cookhouse, under the cook wagon or to the side in a tent. Wake-up time was usually three a.m. so breakfast could be ready by five.

The cooks baked batches of bread two or three times a day. It took that much for the five meals they served. These included the three normal meals plus a lunch of sandwiches and coffee in the forenoon and afternoon.

There were times when the cook wagon had to be moved just before a meal much to the displeasure of the women. It meant dishes had to be packed on the floor, the water barrel covered and anything loose had to be secured for a rough ride.

It happened once in the Palouse country that such a move was necessary just before a meal. The boss hitched up four horses. Two of them were spirited three-year-olds and figured the fastest way to get the job done was at a dead gallop—downhill. Although the boss had a firm grip on the reins, had his feet securely braced on each side of the cookhouse door and the brake in the last notch, there was no stopping the horses as they flew down the slope. Finally pulling them up at the bottom of the hill he looked back to see every cupboard door and drawer open with two wide-eyed cooks sitting in the midst of pots, pans and kettles.

In the June 29, 1947 issue of the Walla Walla Union Bulletin, Mrs. Cora J. Livingston Kerns reminisced about the early harvest days. It's obvious in reading her story that a cook wagon was not used in her father's setup, but there was still food to provide.

In spite of the hard work there was plenty of

WOMEN WERE HARVEST HEROES TOO. The gals in the doorway worked easily as hard or maybe harder than any of the crewmen gathered around the cookhouse. This harvest scene took place near Ritzville, Washington. Left to right the two boys standing on the edge of the wagon were Henry Hinemann and Henry Nauditt. Standing on the ground second from the left was John Bauman; third, Ed Freise; fourth, Antone Nissen; seventh, Roy Pellner; ninth, Joe Tishner. The fellows seated on the ground were unidentified, but they were a couple of clowns, one with a basket from the kitchen over his head.

Courtesy, A. M. Kendrick

reason for young girls to look forward to harvest time. As the story headline stated, the "early-day wheat harvest was not without its romantic side."

In talking about those times she pointed out that men weren't the only hard workers when she said, "The women had an arduous time of it too. All would be bustle and hustle around the big kitchen for days before the crew arrived. Big tables had to be set up, with benches down the side to serve from 15 to 17 men. Bread, cakes, cookies, pies, hams and much other good, solid food was prepared.

"Always there were two hired girls to help mother and they usually went with the crew from house to house. As did the men, each one carried her own bedroll. Neighbor boys would bunk together in the hay-mow. Sometimes they would find burrs of weeds in their beds when they wearily lay down to sleep.

Those hired girls had found time to do something besides cooking. The boys usually found out some way to get even in a return prank.

"The men made their toilet down by the spring. No hot water, but plenty of cold water in basins, with lots of soap and towels. A looking glass would hang in a tree, with a comb or two nearby.

"These boys made a neat appearance at meals. The hired girls who served them were local and well known, and just as attractive as any nowadays. It was a romantic opportunity, cooking for a thresher. Youth had the same reaction then as now, but was more bashful.

"Cooks arose at 3 a.m. to serve breakfast before dawn, as the boys had to be in the field by daybreak. The girls retired no sooner than 10 or 11 p.m., so generally managed an afternoon nap.

FAMILY SHOT. The Blue family gathered around the wheat-harvest cookhouse for this family shot. Their ranch was located south of Moro, Oregon.
Courtesy, Sherman County Historical Society

"However, this nap was sometimes dispensed with so the girls could visit the field, where they took a ride around with their favorite header-box driver on his way out and back to the thresher. Some near spills on steep hillsides would furnish excitement for the trip but never were there any real casualties. In fact in all those years of operations no one was badly injured."

MEALTIME. It wasn't uncommon for the animals to eat first as in this scene. Painted on the side of the cookhouse was the name A. J. Douglass. He was a commercial thresherman in Sherman County, Oregon during the early 1900's who provided harvesting services for wheat ranchers who didn't have threshing facilities of their own.
Courtesy, Sherman County Historical Society

COOKHOUSE QUEENS. Food fit for a king came from cookhouses such as these. The equipment the ladies had to work with was simple but the food was delicious according to the crews. The bench to the lower left with the wash basins was the washup area for the crew as they came in for meals. Towels and a mirror hung on the side of the wagon. Louise and Rebecca Miller somehow managed to look young and pretty in spite of the long hours of work they put in providing food for a hungry crew near Ritzville, Washington.

Courtesy, A. M. Kendrick

COMEDIANS BUT NO LAUGHS. It's frequently said that there's a ham in every crowd. There were four in this one. One had a tub he used as a drum. A second strummed an axe while two others played the young lovers. Surprisingly, such antics didn't evoke one single smile.

Courtesy, Sherman County Historical Society

CREW, THRESHER AND SLING. The sling was left suspended in the air when this shot was taken on the George Stelman ranch near Nez Perce, Idaho in 1908.

Courtesy, Ted Worrall

PICTURE GALLERY OF HARVEST HANDS. A Buffalo Pitts steam engine formed a pictorial backdrop for the crew on the Charles Tom ranch of Rufus, Oregon. Grandson Allan Tom became a leader in the wheat industry during the 50's.

Courtesy, Sherman County Historical Society

IRONSIDES AGITATOR SEPARATOR

BUILT TO THRESH, CLEAN AND SAVE GRAIN.

24 TO 40 INCH CYLINDERS

BELT MACHINE   RIGHT SIDE

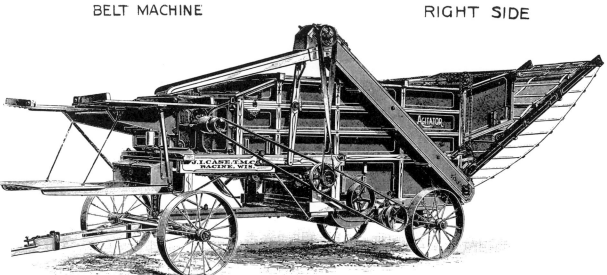

The Ironsides Agitator, as illustrated, is our regular belt machine, fitted with folding feed tables, feeder's foot board, tailing elevator, tailer, straw stacker (folded) and belts.

The Feed Tables are simply folded over hopper, in moving. No time wasted by having to remove them.

The Band Cutter's Stand has adjustable legs and when attached to bolster brace the band cutter faces the feeder. When stand is attached to feeder's foot board the band cutter stands beside the feeder.

The Cylinders of the 32, 36 and 40 inch cylinder **Belt** Agitators have 12 double bars, thus making a very powerful cylinder. Double bar cylinders **will not** be furnished with **Geared** Agitators unless specially ordered. In construction they are perfectly balanced on steel shafts, on which all pulleys are placed outside of bearings; the double bars, which are strongly banded, add strength and weight to cylinder, also allow of long, tapered shank to teeth which greatly strengthens them, and the teeth are held more securely. The cylinders are hung between the Ironsides which form rigid bearings for the extra long cylinder boxes.

The Cylinder Boxes have our new patent compression grease cup; this is made as part of the box and is intended for Helmet oil or other hard oil and insures the journal running cool. In addition to this the lower half of box has a chamber for oil; which is for use of ordinary oil when hard oil is not at hand.

IRONSIDES AGITATOR. In the late 1800's Case built its reputation in the threshing world around the Ironsides Agitator Separator. This is how it was pictured in the 1897 Case catalogue. At the tail end was a web-type straw stacker. These were common and widely used until the wind stacker was introduced. It blew the straw away from the separator.

Courtesy, J. I. Case Co.

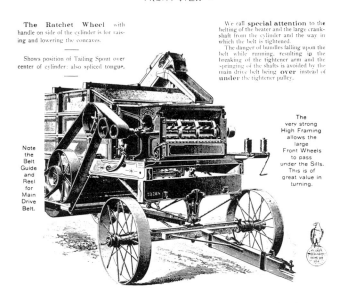

IRONSIDES AGITATOR
FRONT VIEW

The Ratchet Wheel with handle on side of the cylinder is for raising and lowering the concaves.

Shows position of Tailing Spout over center of cylinder; also spliced tongue.

Note the Belt Guide and Reel for Main Drive Belt.

We call special attention to the belting of the beater and the large crank-shaft from the cylinder and the way in which the belt is tightened.
The danger of bundles falling upon the belt while running, resulting in the breaking of the tightener arm and the springing of the shafts is avoided by the main drive belt being over instead of under the tightener pulley.

The very strong High Framing allows the large Front Wheels to pass under the Sills. This is of great value in turning.

Notice our Ball and Socket arrangement for front axle. This device adds long life to the machine, as it prevents the rocking, the jarring, and the pounding of the Separator, while being moved along rough roads. It insures freedom and easiness of motion. The front end is always level with the rear end. In setting, it is only necessary to level the back axle, which is a great advantage. This feature in the mounting of Separators has been tested for years—and is far superior to the old fashioned bolster plate.

FRONT VIEW. There wasn't hardly any angle or portion of the Ironsides Agitator threshing machine that wasn't shown in the 1897 Case catalogue. This wood engraving features in full detail the machine's front end arrangement.

Courtesy, J. I. Case Co.

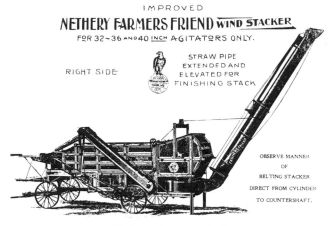

IMPROVED
NETHERY FARMERS FRIEND WIND STACKER
FOR 32~36 AND 40 INCH AGITATORS ONLY.

RIGHT SIDE

STRAW PIPE EXTENDED AND ELEVATED FOR FINISHING STACK

OBSERVE MANNER OF BELTING STACKER DIRECT FROM CYLINDER TO COUNTERSHAFT.

The best testimony concerning the popularity of wind stackers is, that they are most used where they are best known.

WIND STACKER ATTACHMENT. The same 1897 Case Ironsides Agitator Separator was also sold with a wind stacker instead of a web-type straw stacker. The acceptance of the wind stacker was almost immediate. The above model featured the horizontal fan which the company claimed best because the straw fell into it by gravity rather than having to be mechanicaly fed into it.

PANORAMIC. It took the old-time panoramic camera to get such an all-encompassing picture such as this.

Courtesy, Washington State University Archives

THE DESIGN OF THE IRONSIDES AGITATOR IS BEST.

SECTIONAL VIEW
OF 1897 MACHINE.

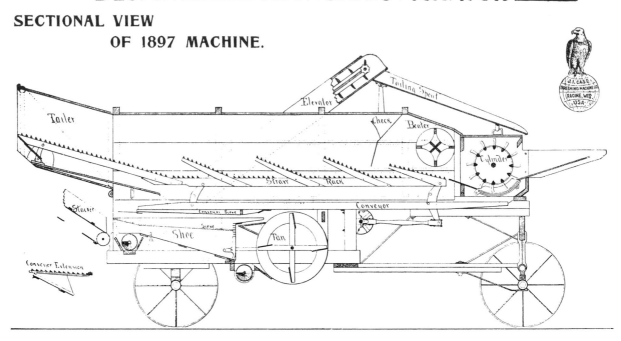

SECTIONAL VIEWS OF SHOE WITH SIEVES IN PLACE FOR DIFFERENT KINDS OF GRAIN.

There are 7 Notches and 5 Truss Rod Holes for Sieves in Shoe.

Shoe with Three Sieves Placed for Weedy Flax and Grass Seeds.

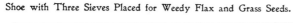

Fine lip wheat sieve, top notch, top hole.
Zinc flax sieve, third notch, third hole.
Screen sieve, sixth or seventh notch, bottom hole.

Shoe with One Sieve, Second Notch, Third Hole.

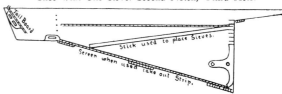

For wheat use fine lip sheet iron sieve.
For oats use large lip sheet iron sieve.
For barley use either oat or wheat sieve.

Shoe with Two Sieves Placed for Flax.

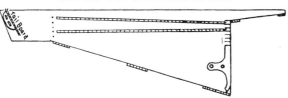

Perforated zinc sieve, top notch, top hole, tail piece attached.
Tail board must be set to suit the work.
Screen sieve, fourth notch, bottom hole.

Shoe with Two Sieves as Set for Wheat.

The two sheet iron sieves are sometimes used for wheat.
Tail piece may be used when elevator carries too much chaff.
Small lip sheet iron sieve, top notch, top hole.
Wire wheat, third notch, fourth hole.

INSIDE LOOK AT SEPARATOR. The 1897 Case catalogue included a cutaway drawing showing the inside workings of their threshing machine. It also featured the adjustments necessary for different kinds of seed to be harvested. The sectional drawing in the lower right-hand corner shows the setting for wheat.

Courtesy, J. I. Case Co.

WORLD CHAMPION WHEAT. John Purdy, 81 years old, sat proudly with a sheaf of wheat that had been judged world winner at the 1900 International Exposition in Paris. He had grown the wheat on his farm. Purdy, founder and developer of Purdy Hot Springs, lived near Athena, Oregon and farmed on Wild Horse Creek. The striking thing about this bunch of wheat is the extreme length of the straw. The wheat plants of today are much shorter. The excessive straw has been bred out of the modern-day wheat varieties.

Courtesy, Paul Walden

BELT POWER AT IDAHO THRESHING SCENE. The lifeline of power for the threshing machine was the long belt looped back to the huge pulley wheel of a gas or steam engine. As long as the belt hummed and danced over the ground the separator hungrily gobbled the headed or bundled wheat as it was pitched into the feeder. This 1906 threshing scene was taken near Nampa, Idaho. The pitchfork man at the feeder was cleaning stray wheat off the ground.

Courtesy, Idaho Historical Society

HARVEST ON THE JOSEPH TUCKER RANCH. Joe Tucker must have been a bit of a genius considering all his many interests and mechanical acomplishments. For one thing he was probably the first wheat-farming student of weather in the Northwest. He not only studied it—he also predicted it. Tucker didn't make his prognostications by just looking at the Farmers Almanac though. His scientific turn of mind inspired him to make a barometer and a wind gauge. These coupled with a keen eye on the sky made him "the" weather oracle of the area.

Joe Tucker also led the way when it came to telephones. It didn't take long once he set his mind on the idea of a community telephone system to have the instruments wired in his own and his neighbors' houses. Local folks put the poles and wire up under his supervision. The main switchboard was in the Tucker home and Mrs. Tucker was the operator.

To top it all, Tucker was a photo bug. He took the above picture, but no one is quite sure how he did it since he is also in the photo. Joe Tucker is the gentleman standing on the threshing machine with the beard. That very evening he developed it and printed it in his own darkroom at the house.

This threshing was done in approximately 1913 on the Tucker ranch in an area known as Tucker Prairie. Three families were the first settlers on the prairie land. They were the Tuckers, Harrisons, and Robinsons. Since the Tucker family arrived there first, it was dubbed Tucker Prairie. It was north of Cheney, Washington. The harvesting outfit was owned jointly by Tucker and C. O. Vaughn. Mr. Vaughn was in the buggy. Mrs. Vaughn was perched on the water wagon in the white hat. An 18 HP return flu Gaar Scott steam engine provided the power.

Courtesy, Mrs. Martin Thiemens

THRESHER TRAIN. On May 10, 1889, a special train pulled into Portland loaded with threshing machines, and steam engines. There were 72 threshers in all. The man responsible for the trainload of harvesting equipment was Z. T. Wright. All machinery came from the Advance factories. For shipping, the separators were not setting on their wheels. The wheels were no doubt loaded flat under the frames to save space. All the threshers were hand fed and featured web-type straw stackers.

Courtesy, Mrs. Harriet Moore, Benton County Pioneer Historical Society

BOTHERSOME DUST. The main dust producer on a threshing machine was the blower. At a new setup, a great deal of thought went into placing the separator so the dust would blow away from the workers. The sky was scanned and the wind was checked to assure the thresher was spotted correctly. Nine times out of ten the advance planning was a waste of time as the wind invariably shifted just as the threshing started. This picture shows a sack-jig teamed with two sack-sewers. One of the sewers had finished lacing the top of his wheat sack and was carrying it to the newly started pile. The other sewer was in the sewing process.

Courtesy, Sherman County Historical Society

COVER SHOT. The American Thresherman magazine featured this shot on the cover of its April, 1920 issue. The harvest was on the George Howell ranch in Albion, Washington in 1908. The steam engine was a 20 HP Gaar Scott.

Courtesy, Ted Worrall

HARD AT WORK. The Deer Flat area south of Caldwell, Idaho was the site of this 1906 stack-threshing scene. The crew was pitching from both sides of the feeder. The only man who had time to mug the camera was the engineer on the steam engine.

Courtesy, Idaho Historical Society

SACKING WHEAT. Sack-filling and sewing for most threshing outfits took place under an improvised shelter. Two men were kept plenty busy with their sacking job on this operation. Usually there were three men doing the sacking —two sack sewers and one sack jig. The ghost-like image of one of these men photographed while carrying a full sack out to the pile indicates they had to move fast to keep up with the wheat kernels that came cascading out the sack spout. It's interesting to observe the style followed in piling the sacks of grain on the ground. The first layer of sacks was set on edge. The reasoning was that less grain became exposed to whatever moisture might be picked up from the ground.

Courtesy, Sherman County Historical Society

STACK ACTION. Pipe smoking seemed to be the popular pastime of at least three of the crew members around this 18 HP Minneapolis steam engine. There's no pipe smoking being done, however, by the men working on the stack. They're too busy for that. While two men pitched the headed wheat onto the feeder a third pulled the Jackson Fork back down the stack. To set the teeth of the fork in the wheat he jammed it down with his foot and the weight of his body. The rope from the Jackson Fork went up through the overhead pulley and out to the team at the far left of the picture. When the forkman yelled, the team was driven forward. This set the fork with its teeth full of wheat in motion. It moved slowly down to the wing feeder. Once in position the forkman dumped the load and started the whole procedure over again by grabbing the fork and pulling it back down the stack.

Courtesy, Francis A. Wood

A NORTHWEST THRESHING MACHINE. The Pride of Washington separator was manufactured by the Gilbert Hunt Company in Walla Walla, Washington. *Kirby Brumfield*

PICTORIAL REMINDER. Herman Curtis was a natural born public relations man. Whenever he threshed a farmer's grain he always had a photo taken as the job was done. He gave a copy to the wheat grower with an inscription showing how fast and efficient his crew was. In this case he and his men took only nine hours to fill 1800 sacks with newly threshed wheat. The farm was in the Clear Creek district some six miles east of Colfax, Washington. Curtis brought his crew and equipment over from his home at Palouse, Washington. The steam engine was a 22 HP Advance straw burner. The engine was already hooked up to the separator and ready to move out. The man on the horse got his laughs for the day by putting his hat on the horse's head.

Courtesy, Mrs. Glen Epler

JACKSON FORK. This picture makes it fairly obvious that working on the stack was no picnic. It was hot, dirty work. The Jackson Fork was used to bring large scoops of wheat from the stack to the threshing table. There, hoe-down men pitched the headed wheat into a feeder leading to the separator.

Courtesy, Idaho Historical Society

SACKING. Fifteen men took time out from threshing to have their picture taken. No one looked particularly happy about the idea. At the left of the picture is a good view of the threshing machine's sacking mechanism. A sack was hooked to each of the two openings. Only one sack was filled at a time. The round stick seen in the middle of the double chute was attached to a divider board inside. When one sack was filled the lever was flipped in the opposite direction. This shot the grain out the second opening into the waiting sack.

Courtesy, Sherman County Historical Society

SWEEP-POWER

The first threshers were powered by horses. Six or seven teams were hitched to arms which extended from a huge gear wheel. The apparatus was called a sweep-power by some. Others referred to it as a horse-power. As the horses walked around and around in a circle, power was transferred to the separator through a tumbling rod.

In southeastern Washington, Mrs. Cora J. Livingston Kern's father used a horse-power for their thresher in the 1890's. In her two stories published June 22 and June 29, 1947 in the Walla Walla Union Bulletin she described how this operation worked along with additional details on the entire harvest.

"The threshing machine and horse-power first operated on Mill creek was owned by R. J. Livingston who settled around 1878. It was with a great deal of childish delight that I used to watch the old tumbling rod, that ran from the power to the separator, turn 'round and 'round. Horses learned to step gingerly over it every time they came around on the power side. Once in a while it would catch on fire due to friction, and cause damage before it was put out.

"The threshing machine was run with a heading outfit. Fifteen to 17 men were required to operate this

outfit; seven on the heading crew and eight with the thresher. Besides the header-driver (or puncher) and four drivers on the grain boxes—who hauled headed grain into the thresher—one man helped load the wagons and one helped unload them. Sometimes on very steep land a man was needed to help elevate the header.

"Everyone was kept busy from daylight to dark to save the heavy crops while the weather held good. Yields commonly ran from 40 to 60 bushels per acre. Winter wheat was the main crop. Much grain was wasted as the headers could not pick up the down grain.

"Eight men ran the separator and comprised the following: John Livingston, who ran the thresher; Chris Miller, who drove the horse-power; two hired men on the feeder; a sack-jig; a sack sewer; two more men to keep the straw away.

"There was no such thing as a blower on the thresher. The straw came tumbling out the back end to pile up nearby. It was a very dirty job for the man who bucked it out to the stack. He usually rode well up on the stack with the load before he yelled 'let go' to the straw cart driver.

"There was chaff flying everywhere. Young boys

82

WOOD ENGRAVING OF SWEEP POWER. Horses were the sole source of power on the farm prior to steam and gas engines. One of the ingenious devices developed to transfer power from the horse to the threshing machine, baler, grist mill or whatever, was called a sweep power. A booklet of wood engravings published by the J. I. Case Co. in 1942 featured several representations of sweep powers. This unique art from a past age depicts 12 horses as they walked around in a seemingly endless circle. The man atop the sweep rotated along with the horses. Going out of the picture to the left was a tumbling rod. This rod was attached to the machine being powered. Each time the horses made a circle they had to step over the rod.

Courtesy, J. I. Case Co.

of the neighborhood always looked forward to the time when they could drive the straw cart and go with the crew of grown-up men. This was the easiest job on the outfit, but not the cleanest.

"Much grain had to be salvaged around the old-fashioned loading platform. This platform opened on all sides, where the headed grain was dumped and forked by hand into the machine. It would get scattered and some would fall off.

"The man responsible for seeing that as much could be saved as possible was told to "Hoe it down," hence the term "hoe down." We are not sure which originated first—the "hoe down" as applied to threshing, or the term used in the old song of earlier days about "the old-fashioned hoe down" as indicative of a dance. But hoe it down they did.

"Some 40 head of horses were necessary. Six were used on the header; 16 on the four header boxes;

12 on the horse-power and two on the straw cart. Then there was the roustabout's team and two or four on the grain wagons.

"We hauled grain to Tracy, then known as 'the dump,' in wagons with a California rack and high spring seats. The main reason, we suspect, for the high seat was so the driver could be above the fog of dust churned up by the horses. It would sometimes be six inches deep in the road in hauling time. The wheels would climb out of one chuckhole only to drop into another—hence the spring seat.

"John Livingston always tended separator, while Chris Miller drove the horse-power. For a week or so John would be very busy preceding harvest, getting the machine in proper shape to run through the season with as little stoppage as possible.

"Rain was about the only thing that caused a lay-off for the outfit. There was very little that could go

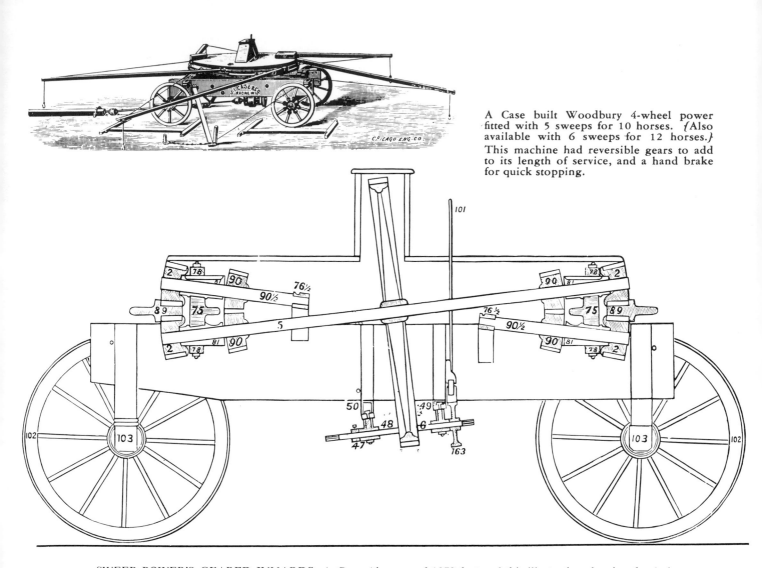

A Case built Woodbury 4-wheel power fitted with 5 sweeps for 10 horses. *(Also available with 6 sweeps for 12 horses.)* This machine had reversible gears to add to its length of service, and a hand brake for quick stopping.

SWEEP POWER'S GEARED INNARDS. A Case Almanac of 1879 featured this illustration showing the design of Woodbury sweep powers. Number 89 is the rotating member to which sweeps were attached to turn the large bull gear and tumbling rod.

Courtesy, J. I. Case Co.

wrong with the horse-power, outside of the coupling rods to the thresher.

"I can see Chris yet, standing on the old power, slowly turning with the platform all day long, swaying his long whip and encouraging the horses by his gentle voice. I never saw him hit a horse and neither did anyone else. The horses were arranged with the fastest on the outside. Chris was a master hand at getting the horses to work evenly and steadily—so the power could drive the separator with maximum efficiency."

Mrs. Kerns went on to explain, "The usual run was around 30 days. From 20 to 30 acres was the usual day's run, which would mean around 600 sacks of grain, or 1,350 bushels. It would seem with so many men employed, to be a costly method, but wages were low. Men got from 75 cents to $1 per day for from 13 to 14 hours. Hired girls received the big sum of 50 cents.

"When the season was over there was the settling up to be done. John Livingston kept the time and I used to get a thrill watching the proceedings. Someone brought the money out from Baker-Boyer bank.

"None of it was paper bills. Father John would pile the gold and silver coins in neat stacks on the front-room table. The men came in one at a time while Chris Miller usually stood in the door. As the men never drew their wages until they finished, they would check up their time, take their share and go out with jingling pockets and happy faces.

"To many, it was a winter's grubstake. To neighbor boys it meant helping the home folks pay the harvesting costs and they were lucky to get a new suit out of it. But all had the satisfaction of a job well done and looked forward to the next season's operations. It was the big event of the farmer's year and a well rewarded community effort."

WHO NEEDS STEAM. A 12-horse rig ran the 36 inch Case threshing machine in the background. The sweep power along with the web-type stacker on the separator were considered primitive equipment in 1916 when the photo was taken. The owner, Willie Herman, apparently figured this arrangement worked fine and stuck with it rather than pay out the two or three thousand dollars needed to buy a steam engine. Al Herman, who supplied the picture, was not related to Willie Herman.

Courtesy, Al Herman

BUSIEST PLACE ON THE RANCH. This operation near Colfax, Washington was likely photographed near or before the turn of the century. The tumbling rod can be seen as it went from the sweep power down to the threshing rig. Even though the wagons standing by are built like header-boxes it appears they are filled with bundled wheat. The small dots on the hill in the distant background also suggest bound grain. Binding wheat in the hilly Palouse country near Colfax would have been slightly irregular since most of the grain of that area was headed rather than bound. Bedrolls in the foreground indicate the men followed the normal practice of sleeping around the straw pile.

Courtesy, Bill Walters

AROUND AND AROUND. Weeks later it was easy to tell where a sweep power was spotted in a horse-driven threshing operation. The horses marked out a very definite circular path as they plodded around and around. There wasn't a spear of grass or stubble left in that trackway. The sweep power is barely visible at the left of this picture.

To the right is a team hitched up to the cook house. The driver was positioned in the doorway when he moved the kitchen-on-wheels from one spot to the next. Ocassionally a team would run away. When that happened it meant only one thing—dishes, pots, pans, benches and sometimes cooks spread all over the inside of the wagon. If horses for some reason got panicky and started running, the sound of the clattering utensils behind didn't tend to calm them down.

Courtesy, Wayne Doty

HORSE-POWERED THRESHING. It's obvious who wrote "Barney" above the figure standing on top of the threshing machine. It was written by "me" and the "me" in this case was Barney Sparrow. Barney farmed near Lind, Washington. His wheat made eight bushels per acre with the price at 84 cents per bushel. That was in 1912. Barney was heading and threshing at the same time. He used horse power to run the separator. Twelve horses and mules did the job. They are seen at the right of the picture. Barney Sparrow was also an auctioneer. In his 40 years of auctioneering, he cried over 2200 farm sales.

Courtesy, Ted Worrall

NO STRAW STACK HERE. The straw from this horse-powered thresher was fed directly into the hayloft of the barn rather than being stacked outside. The separator was a 32 inch Case Eclipse. It was purchased in 1888 by a gentleman named Croissant of Lyons, Oregon. This early Eclipse was a hand-fed thresher.

Courtesy, Al Herman

STRANGE SETUP. With seven arms it would seem this sweep power was powered by 14 horses—not so. It looks more like 17 horses. Close inspection reveals several hookups had three horses teamed together rather than the normal two horses. The photographer clicked the shutter for this shot in 1904 near LaCrosse, Washington.

It's interesting to observe how the tumbling rod went from ground level where the horses stepped over it to approximately three feet above the ground as it hooked into the thresher. Off the left shoulder of the man standing on the ground is a tripod affair which supports the elevated rod.

The pile on the left was headed wheat to be threshed. The pile on the right was straw. The threshing machine delivered the straw to the stack with a web-type stacker.

Courtesy, Charles Freeburg

STEAM POWER

Steam engine! Say those two words to anyone who ever worked around one of the gentle monsters and you better be ready to talk a spell. Nothing will spark a flow of memories quite like those two magical words. Observing a steam engine in action was a sight that didn't fade from a person's mind. These gentle giants were a feast not only to the eyes, but to the ears as well.

To the eyes it meant a huge, gear-laden engine, throwing a plume of black smoke and white steam into the warm summer sky; a long belt flopping and dancing just above the wheat stubble as it made one round trip after another from the steam engine to the thresher; and an engineer, fireman and water tankee who tended to the friendly monsters every need.

The engines tingled the ear with their hissing and huffing. The threshing crew pitched, sacked and stacked to the accompaniment of the exhaust's soft, rhythmic chug-a-chug. The piercing shrill of the steam whistle said more than words when lunch and quitting time rolled around.

It was in the late 1880's up through the turn of the century when steam engines began moving in on wheat harvesting in the northwest. These huge powerhouses quickly replaced horse-power threshing.

There was nothing agile about the engines. They were heavy and slow-moving, weighing as much as 15 to 25 tons apiece. They had power though, the likes of which farmers had never seen before. The steam engines sent a quiet strength humming over the drive belt that got the job done quicker.

These steam-powered iron servants came in a great variety of sizes and shapes. Their horsepower ranged from 6 to 150. One of the more popular models was the Case 110. Advertisements told of a simple engine, a locomotive type boiler, cab and power steering.

It's amazing how this behemoth could pull the 16 plows it frequently did considering the size of its iron wheels alone. They were seven feet high. When the extension rims were added to create greater traction they measured more than four feet wide and weighed in excess of two tons each.

With wheels so large and heavy it seems all the engine power would have been taken in just turning them. The entire tractor scaled out at 42,000 pounds and cost $4,200 or roughly ten cents a pound.

A steam tractor's power was designated by two numbers. One was the drawbar rating, the other was the belt rating. An engine which was rated at 16-48 meant it supplied 16 horsepower on the drawbar and 48 horsepower on the belt. In this case, the drawbar rating indicated it could pull as much as a sixteen-horse team could pull.

The largest Case engine in general use was known

FAMOUS OLD CASE. This is the famous Case No. 1. As with all early steam engines it provided belt power but could not motivate itself. To move the engine from one setup to the next required a team to pull it. The seat on the smokestack looks curiously out of place in the picture. During a move though, the stack hinged back into the U-shaped cradle and the man driving the team had a convenient place to sit. This engine was built around 1870.

Courtesy, J. I. Case Co.

as the "110' because it developed 110 horsepower on the belt pulley.

The ratings varied with each manufacturer. Many will argue they were underrated when compared with today's gasoline or diesel powered tractors.

Although there were many different kinds of steam engines mounted on these tractors they all had one basic similarity. Compared to modern motors they turned very slowly. They averaged only 250 to 300 revolutions per minute. This one characteristic led to many years of life and dependable service from the steam powerhouses.

Interestingly, these old-timers had no gear boxes. The result was one speed and one speed only. They ran roughly the same speed loaded as they did empty. For most steam tractors that speed was a blazing three miles an hour with governors wide open.

Head man on the steamer was the engineer. He was held in special esteem by the community as he piloted his ponderous engine through the wheat harvest.

Drivers took quiet pride in their ability to have a thresher operating in record time at a new setup. It meant lining the engine up with the separator. This was no small job. The steam-plant-on-wheels was frequently as far as 60 feet from the thresher. The alignment had to be perfect or the revolving belt would jump off the pulley. Many of the more experienced engineers could eyeball their machines into position on the first try.

As soon as the engineer had his engine in position, one of the crewmen looped the drive belt over the engine pulley and the other end around the thresher pulley. That was the engineer's cue to move the iron giant backward until the belt tightened.

There was no way to disengage the pulley. It always ran. The belt was set in motion as the steam engine backed up and put tension on it. If everything was in line the engineer opened the throttle and it was back to work for the entire crew.

According to the fellows who were there, that's when the puffing and hissing of the steam engine, the slapping of the belt and the whirr of gears blended into a symphony of beautiful sounds.

The engineer wasn't necessarily the first man out to the engine in the morning. That dubious honor usually fell on the fireman. By 4:00 a.m. he was walking through the crisp stillness of an early morning headed out to the slumbering iron hulk. Once there he cleaned the flues, laid the fire, kindled it and nursed the engine along until it built up a head of steam.

By 5:00 a.m. it was midway between night and day. Glimmers of light were warming up the horizon. The eerie morning was suddenly shattered by a piercing blast from the steam whistle. That was the signal for breakfast. At 6:00 a.m. the engineer opened the throttle and the threshing crew swung into action.

One other man tended the steam engine. He was the tankee, sometimes called the water flunky. It was up to him to keep a large supply of water on hand.

At best the steam tractor proved to be an extremely thirsty worker. The tankee had to fill the water wagon three to four times a day. If he had more than three miles to go for water it was a real strain to get back to the steamer before it ran dry. The engineer never let the tankee forget how urgently the water was needed. The reminders came in the form of toots on the steam whistle.

The water-tank driver got a workout every time he pumped the water wagon full. The wagons held from 300 to 450 gallons. Actually, water tank ca-

POPULARITY OF CASE. The Case Co. sold more steam engines than any other manufacturer. This was an 11 x 11 inch cylinder, simple 25 HP traction engine taken from a page of the 1906 catalog. An engine that could move itself about was called a traction engine.

Courtesy, J. I. Case Co.

12 HP Advance

22 HP Minneapolis

STEAM COLLECTION. Steam engines abound at the Paulson Park Museum in Pendleton, Oregon. Ed Paulson has on display numerous steam engines and several threshing machines as reminders of how it used to be done.

Courtesy, Ed Paulson

6 HP Russell

20 HP Case

6 HP Case Portable

13 HP Gaar Scott

REMINGTON OR BEST? This photo generates considerable discussion among steam buffs as to whether the engine above is a Remington or a Best. Remington originally designed this type of steam engine at his small iron works in Woodburn, Oregon. The Remington approach to steam engines featured three wheels instead of four, the vertical boiler, and the main weight of the boiler and engine directly over the two drive wheels rather than distributed more evenly between the front and rear wheels. This resulted in more traction and greater ease of steering. The added traction came from the bulk of the weight sitting over the rear wheels. The Remington approach took the strain out of steering by taking the weight off the front end and by originating a very simple turning mechanism in the single front wheel. The Best Co. of San Leandro, California thought the new approach had considerable merit and bought the rights from Remington. The first models from Best were virtually duplicates of the original Remington. That explains the puzzlement as to whether the pictured engine is a Remington or Best. It's generally believed to be a Best though, since so few Remingtons went east of the Cascades. The scene was taken in the Pendleton, Oregon area.

Courtesy, Buz Howdyshell

STEAM BY BEST. Another Best engine found its way into Pendleton wheat fields as is obvious in this Charles Moore photo. Note the width of the wheels.

Courtesy, Paul L. Kertesz

pacity was figured in "barrels." According to this terminology the water wagons commonly held 10, 12, or 15 barrels.

These mighty steamers were not only thirsty but also had ravenous appetites. Providing the food was the fireman. If the engine was a straw-burner it meant steadily pitching straw into the firebox. Otherwise, it entailed spoon feeding the throbbing powerhouse with wood, occasionally coal.

Associating the smell of burning wood with machinery would be strange by today's standards but was taken for granted then.

91

REAR VIEW OF OUR TRACTION ENGINE.

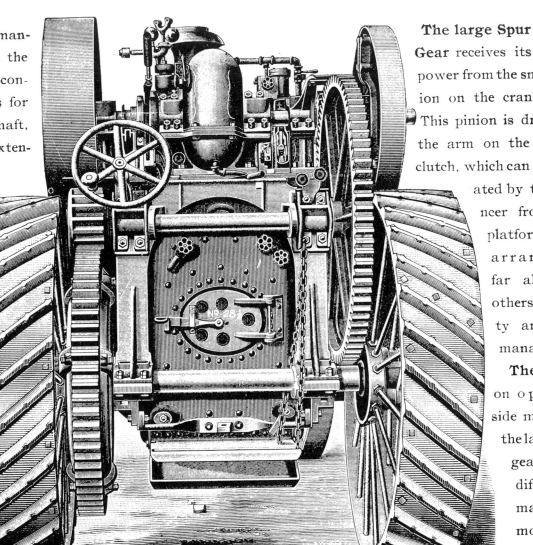

Shows the manner in which the gear frame brackets, containing the bearings for **rear** axle and cross shaft, **are** attached to the extensions on the boiler.

Note the arrangement of cross shaft crossing the boiler above the fire door, held in place by strong babbitted boxes, and by having the gears on either end of shaft, close to the bearings, the driving strain is equalized.

The large Spur Gear receives its power from the small pinion on the crank shaft. This pinion is driven by the arm on the **friction** clutch, which can be operated by the engineer from the platform, by an arrangement far ahead of others for safety and easy management.

The Pinion on opposite side meshes in the large spur gear of the differential, making a most powerful drive.

Shows the manner in which the gear frame castings containing the bearings for rear axle and cross shaft are attached to the extension on the boiler.

SEE THE FOUR BEARINGS OF THE MAIN CRANK SHAFT.

CATALOGUE PICTURE AND PROSE. A rear view of one of many illustrations in the fifty-fifth annual catalogue published by the J. I. Case Co.
Courtesy, J. I. Case Co.

STRAW-BURNER. Steam engines were commonly fed wood in the west. Coal did the job in the midwest and east. This friendly giant, however, performed on a steady diet of straw. Obviously, therefore, it was called a straw-burner. The straw was pitched into the funnel-like opening on the second level in the rear of the machine. It took a steady, continuous motion so that straw was burning in the firebox all the time. It wasn't hard work but according to steam enthusiast Jeff Richardson, it didn't leave much time to wipe your nose. This harvest scene took place near LaCrosse, Washington.

Courtesy, Charles Freeburg

HORSES AND STEAM. The George E. Wood threshing outfit takes a short breather from the hard work of the day during a harvest of about 1900. The frilly-topped steam engine was a Minneapolis.

Courtesy, Francis A. Wood

MAN AND IRON BEAST. Tom Brock stood next to his huge Twin City 40-65 steam engine near Starbuck, Washington in 1921.

Courtesy, Ted Worrall

THE BEST. The people of Rexburg, Idaho won't soon forget how a Best steamer looked. This 110 HP powerhouse is on permanent display in the city park.

Courtesy, Al Herman

94

NO BELT NEEDED. The shaft extending from the Case steam engine to the threshing machine sets this operation apart from others. Almost without fail, a steam engine powered a separator by a long drive belt. In this case, however, the operator used a drive rod that could be loosely compared to today's power take-off on tractors. The shaft was very likely the tumbling rod from a discarded sweep power.

The engine shown was a return-flue Case built before 1900. This unique approach to providing power to a threshing machine took place in the Pendleton, Oregon area.

Courtesy, Buz Howdyshell

WHEAT FIELD GIANTS. The men standing beside the tricycle-wheeled Best steam engine were completely dwarfed by its immense size. The combine too must have been one of largest on record. The sack chute was long enough to hold at least 10 sacks, twice as many as normal. It took a tall ladder just to get the sack-sewer and sack-jig up to the sacking platform. This giant harvest team was working near Rexburg, Idaho in the upper Snake River Valley.

*Courtesy, Idaho
Historical Society*

COOL WATER. Most water wagons used in the wheat harvests of the Northwest had this unique cigar shape. The water hauler was kept busy supplying liquid for not only the steam engine, but also for men and horses.

Courtesy, Paul L. Kertesz

SLOW PROCCESION. It's an exaggeration to say, "You had to set stakes to see the forward progress of an outfit like this one." One thing is sure. This combination of a Peerless steam engine pulling a Harris combine was setting no overland speed records. In spite of that, considerable grain was being harvested each day thanks to the 25 foot width of the header. The harvest was near Moccasin, Montana in 1912. The procession included one of the early bulk outfits. The wagon was attached to the harvester and followed alongside. The long tube leading down from the combine delivered the wheat kernels into the bulk wagon and the man standing in the box leveled the load.

Courtesy, Ted Worrall

WATER WAGON. There were no automatic pumps to transfer the water from the wagons to the steam engines. The simple, man-powered pumps usually sat on top of the wagon as in this picture. It was up to the "tankee" or water wagon driver to manipulate the long handle back and forth both to fill the four-wheeled tank and then to switch the water from there to the steam engine barrels. His only rest came as he sat in the driver's seat going back and forth between the steam engine and the water hole.

Courtesy, Sherman County Historical Society

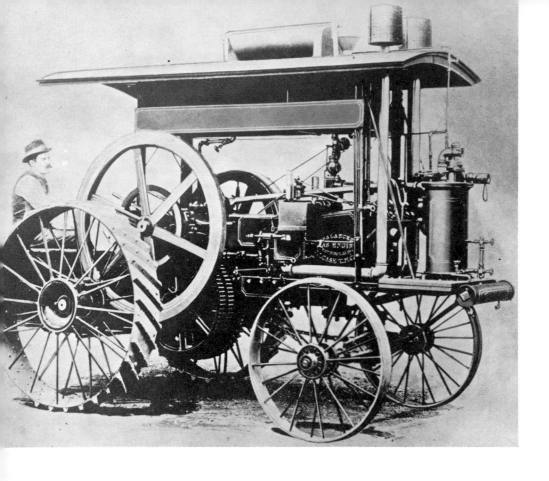

CASE TRIES GAS. The first Case gas tractor was built in 1892.

Courtesy, J. I. Case Co.

OIL PULL. The popularity of steam wasn't easy to overcome. Gas engines made steady inroads into steam engine sales, though, from about 1910 on. This powerplant was a Rumely Oil Pull. It was rated at 20-40 HP. The first gas engines were mammoth just as the steam engines had been. Although it looks large, the Oil Pull was smaller than many of its steam-powered counterparts.

Courtesy, Washington State Historical Society

SIGN ON THE DOTTED LINE. It used to be that the Oregon State Fair provided an area for various tractor companies to demonstrate the plowing ability of their machines. In this shot taken about 1918 the R. M. Wade Co. is demonstrating two of its Heider tractors. They were good enough to prompt C. H. Ernst of Ernst Hardware, St. Paul, Oregon to sign an order for four of them. He was the man with his foot on the plow.

Note the special rocker lugs on the 9-16 tractor. Both machines also featured a very "modern" lift plow. The Heider tractor was one of many hitting the market at this time as the small and medium sized dual purpose gas tractors gained in popularity. Heider tractors were built with seven speeds forward and seven speeds in reverse. Another distinct feature was the belt pulley drive. It could be run seven speeds in either direction, and the pulley could be mounted on either side of the tractor.

Carl Kirsch remembers his father, F. W. Kirsch, bought one of the 12-20 Heider tractors for $2400. The young boy on the tractor was Wade Newbegin, grandson, later president and general manager of R. M. Wade and Co. The company was founded in 1865.

Courtesy, Carl Kirsch

INTERNATIONAL PRO-
DUCES GAS TRACTORS. In
1906, International Harvester
company started regular production on gas tractors like this
one. They were built under an
arrangement with the Ohio Manufacturing Company of San-
dusky, Ohio, which supplied the
truck and transmission on which
an International engine made at
Milwaukee Works was mounted.
The first all-International tractors came out in 1908.

*Courtesy, International
Harvester Company*

GAS ENGINE. Even at the turn of the century there were some threshing machines being powered by gas engines.
Doing the job in this 1900 harvest scene is an International Thermoil engine. It was rated between 15 and 20 HP.
The engines were far from taking over though, as this picture included 11 horses, 19 men and a dog.

Courtesy, Charles Freeburg

AJAX TRACTOR. A gas tractor was manufactured in Portland, Oregon between 1912 and 1914. It was called the Ajax. The number 1 model is shown in the picture. The caption accompanying the picture said, "This is our No. 1 Ajax Tractor, which gives a good idea of its appearance and ruggedness. It develops 40 brake horsepower, weight 16,000 pounds, has from 5,000 to 6,000 pounds pull on the drawbar. Size motor is 6 inch bore by 8 inch stroke, of the horizontal four-cylinder double-opposed type, which is best adapted for traction engine use. We also build a No. 2 Ajax of 70 brake horsepower, mounted with a motor of 7 inch bore by 10 inch stroke, weight 26,000 pounds." As late as 1946 reports had it that there was an Ajax tractor still in operation in the Corvallis, Oregon area. It was rated at 25-45 HP.

Courtesy, Al Herman

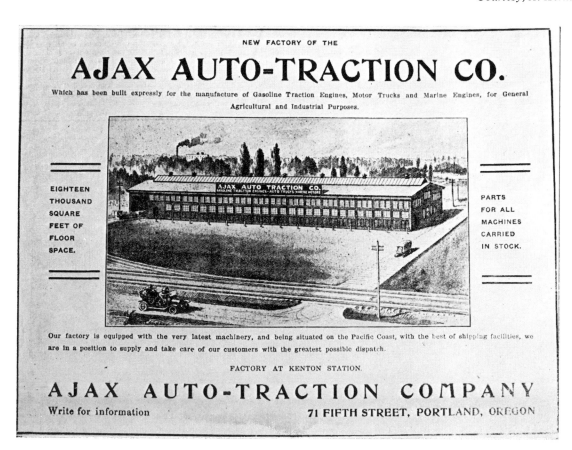

HORSELESS WHEAT BINDER. The Allis-Chalmers Company motorized the six foot McCormick grain binder with their Model 6-12 general purpose tractor. This 1919 version is about to round a corner in a flat grain field somewhere in the midwest. This type of arrangement never gained widespread popularity. It was particularly impractical in the hilly wheatlands of the Northwest.

Courtesy, Allis-Chalmers Co.

MOGUL. International Harvester Mogul 12-25 Tractor, 1913 to 1919.

Courtesy, International Harvester Co.

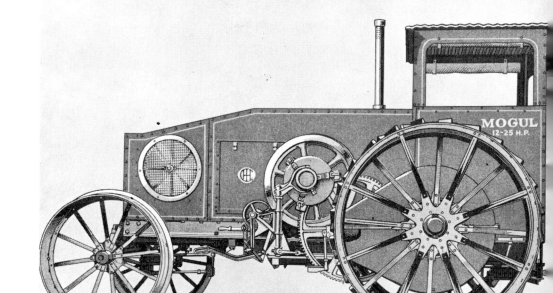

Directory of Manufacturers of Combines, Tractors and Combine Accessories

Company	Address	Product
Advance-Rumely Thresher Co.	La Porte, Ind.	Combines, Tractors
Apex Manufacturing Co.	St. Paul, Minn.	Recleaners
Avery Power Machinery Co.	Peoria, Ill.	Combines, Tractors
A. D. Baker Co.	Swanton, Ohio	Tractors
Buda Motor Co.	Harvey, Ill.	Motors
Carter Mayhew Manufacturing Co.	Minneapolis, Minn.	Recleaners
J. I. Case Threshing Machine Co.	Racine, Wis.	Combines, Tractors
Caterpillar Tractor Co.	San Leandro, Cal.	Combines, Tractors
Charles Closz Co.	Webster City, Iowa	Sieves
Cleveland Tractor Co.	Cleveland, Ohio	Tractors
Climax Engineering Co.	Clinton, Iowa	Motors
Continental Motor Co.	Muskegon, Mich.	Motors
Deere & Co.	Moline, Ill.	Combines, Tractors
Eagle Manufacturing Co.	Appleton, Wis.	Tractors
Electric Wheel Co.	Quincy, Ill.	Tractors, Wheels
Four Drive Tractor Co.	Big Rapids, Mich.	Tractors
French & Hecht	Davenport, Iowa	Wheels
Gleaner Combine Harvester Corp.	Independence, Mo.	Combines
Gray Tractor Co.	Minneapolis, Minn.	Tractors
Harris Manufacturing Co.	Stockton, Cal.	Combines
Hart Grain Weigher Co.	Peoria, Ill.	Combine Registers
Hart Parr Co.	Charles City, Iowa	Tractors
Hercules Motors Corp.	Canton, Ohio	Motors
Huber Manufacturing Co.	Marion, Ohio	Tractors
International Harvester Co.	Chicago, Ill.	Combines, Tractors
Keck-Gonnerman Co.	Mt. Vernon, Ind.	Tractors
Lycoming Manufacturing Co.	Williamsport, Pa.	Motors
John Lauson Manufacturing Co.	New Holstein, Wis.	Tractors
J. T. Tractor Co.	Cleveland, Ohio	Tractors
Le Roi Co.	Milwaukee, Wis.	Motors
Massey-Harris Co.	Racine, Wis.	Combines, Tractors
Mead-Morrison Manufacturing Co.	East Boston, Pa.	Tractors
Minneapolis Steel & Machinery Co.	Minneapolis, Minn.	Tractors
Minneapolis Threshing Machine Co.	Hopkins, Minn.	Combines, Tractors
Monarch Tractor Co.	Springfield, Ill.	Tractors
Nichols & Shepard Co.	Battle Creek, Mich.	Combines, Tractors
Rock Island Plow Co.	Rock Island, Ill.	Tractors
Self-Kleen Thresher Screen Co.	Minneapolis, Minn.	Recleaners
Waukesha Motor Co.	Waukesha, Wis.	Motors
Wetmore Tractor Co.	Sioux City, Iowa	Tractors
Wisconsin Motor Co.	Milwaukee, Wis.	Motors
Wood Brothers Thresher Co.	Des Moines, Iowa	Combines

This list does not include the manufacturers of parts but only those who manufacture complete units.

The 1929 Combine Year Book published this list of leading combine companies serving the wheat harvest industry of the United States.

ADS PLUG VERSATILE TRACTORS. Gas tractors peaked in popularity as they became smaller in size and more practical for field use. This ad was run in the Oregon Farmer during the early '20's.

WILLING WORKERS. From plowing to harvesting, Barney Sparrow's Holt 75 was ready to go. It didn't move very fast but it had power to spare. A motor on the combine ran the threshing mechanism. The tractor was equipped with a large headlight for night work.

Courtesy, Ted Worrall

THE EVOLUTION OF THE BEST

Years 1918-20 25 H. P.

Years 1915-19 40 H. P.

Years 1916 16 H. P.

Years 1914-15 30 H. P.

The "Thirty" of Today

The "Sixty" of Today

Years 1912-19 75 H. P.

Years 1916-18 90 H. P.

Years 1911-13 80 H. P.

Years 1910-13 60 H. P.

THE International success of BEST Tracklayer Tractors is the result of concentration on a high ideal.

The inventive mind of Mr. C. L. Best is responsible for the basic design and improvements which have from time to time during many years past been built into BEST Tractors. He has given constant thought to make Bests better, and each model has been a leader in its day.

In his father's factory, operated from 1886 to 1908 as the Best Manufacturing Company, Mr. Best helped build the sturdy steam tractors of yesterday. The first suc-

cessful tractor of that type for farm use was built by the Best Manufacturing Company, and loaded on cars at San Leandro, February 8, 1889.

In March, 1910, Mr. Best organized the Company which bears his name, and has surrounded himself with an organization of tractor experts, combining large experience with an earnest desire to build the best tractors possible.

The consistent adherence to a correct principle and the steady progress in maintaining this has culminated in the BEST "Sixty" and BEST "Thirty" of today.

Best Steam Tractors of Olden Times

1889 50 H. P.

1890 50 H. P.

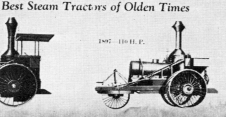

1897 110 H. P.

1900 110 H. P.

THE BEST GETS BETTER. Printed in 1924, this was virtually a family tree showing the earliest and "latest" of tractors from the Best Company of San Leandro, California. The next year the Best and Holt companies merged and formed the Caterpillar Co.

Courtesy, F. Hal Higgins, University of California at Davis

TWO KINDS OF PLOW POWER. It was about 1914 that the contrast between the old and the new was easily observed on the Barney Sparrow farm. The plowing that year was done by both mules and a Holt 75. As time passed the horses and mules in wheatlands across the country were phased out and machine power gradually took over the difficult work. Although this Holt tractor represented the future in 1914 it is an antique from the past by today's standards. It would have to be considered a highly successful tractor in that it spawned a whole new family of crawler tractors.

Courtesy, Ted Worrall

...ing 1914 or later

Uniform Cooling Keeps The Engine on the Job

THE cooling system of a tractor gets its real test during the heat of harvest and early fall plowing. It is then that you can rely upon the Waterloo Boy— the pump, fan and radiator system of cooling always keeps the engine on the job.

A centrifugal pump, four-blade fan, and large size, honey-comb type radiator insure positive cooling on the Waterloo Boy.

WATERLOO BOY
BURNS KEROSENE COMPLETELY

John Deere Implements, Waterloo Boy Tractors and Kerosene Engines are distributed from all important Trading Centers. Sold by John Deere Dealers everywhere.

To secure uniform power you must have uniform cooling. The pump, fan and radiator system used on the Waterloo Boy positively assures uniformity in circulating cooling water.

It holds the engine at the right temperature for proper lubrication, and maintains sufficient heat to insure complete combustion and full power from the fuel.

An even temperature is maintained at all operating speeds because the speed of the pump and fan is automatically controlled by the speed of the engine.

You get a big radiator on the Waterloo Boy. It holds thirteen gallons. You won't find it necessary to stop in the field every few hours on a hot day and fill it.

The cooling system is but one of the Waterloo Boy's superior features. Its simplicity and accessibility, its powerful 12-25 H.P. engine, its ability to burn kerosene and burn it right, its Hyatt roller bearings that eliminate friction, and a drawbar shift lever that gives you the correct line of draft on all tools, all contribute to make it an ideal farm tractor.

We want you to read a booklet describing the Waterloo Boy. Write for it today. Address John Deere, Moline, Illinois, and ask for Booklet WB-535.

JOHN DEERE
THE TRADE MARK OF QUALITY MADE FAMOUS BY GOOD IMPLEMENTS

HOLT 30. Alphonse LeBrun looked as tough and durable as the Holt 30 tractor he stood beside. Belonging to his father, the iron power-plant was Caterpillar No. 5050. It was supplying the muscle for a threshing job in 1914. This Holt Cat was also used to pull four fourteen inch moldboard plows. It did most of its farm work near St. Paul, Oregon. One spring, while plowing through a buffalo wallow, the outfit mired down. It took four men and a stump puller a week to dig it out of the muck.

Courtesy, Carl Kirsch

FROM PLOWS TO TRACTORS. After developing the first successful steel plow in 1837 John Deere plunged into the farm implement business. Within ten years his factory was producing one thousand plows a year. In 1846 he shifted his facilities to Moline, Illinois. The company grew and prospered and became one of the giants in the business of farm machinery. The Waterloo Boy was one the gas tractors that added to the prestige of the John Deere name.

Kirby Brumfield

QUARTET OF CASES. The Case tractor company built and sold the most steam engines in the United States. They followed that era with a popular series of gas tractors that sold well too. This Case dealer displayed his line-up of tractors on mainstreet during a community celebration.

Courtesy, Sherman County Historical Society

IRON COLOSSUS. George Hilderbrand of Wasco, Oregon owned this wheat field giant. Designated a 40-80 Hart-Parr, its power was tremendously underrated. Old-timers say these engines could deliver 100 horsepower with no strain. The tractor had two major drawbacks. One was its own ponderous weight—17½ tons. Most of the engine's power was used in just moving itself. As the picture shows, Hilderbrand plowed down the stubble with his behemoth. It worked fine on the level but was just too heavy and cumbersome to negotiate even the slightest slope.

The second drawback was in the steering. They were extremely difficult to turn. As one farmer said, "It took two strong men and three boys to get the thing around a corner." That was stretching a point admittedly, but there's no arguing the fact it took an experienced hand to do the driving. George Hilderbrand's grandson, Gordon, brings out an interesting fact in pointing up the steering difficulties. "When plowing," he commented, "the driver started by going to the center of the field. He then sunk the plows and went around and around in circles, plowing from the inside out. This eliminated having to turn sharp corners."

Technical data on the machine reveals it had four 9 x 13 inch cylinders set in opposed style. It also featured twin radiators and gas and distillate tanks. The Hilderbrand tractor had extension rims on the wheels which are quite obvious in the photo.

Riding the plows was Val Workman. He was in charge of adjusting the depths of the twelve 16 inch plows. Each of the upright handles controlled two bottoms.

A broken ring gear in this tractor set the stage for one of the most bizarre series of bets in western wheat history. The full story is detailed on the following page.

Courtesy, Oregon State University Archives

MIGHTY MITE. Compared to the size of the Hart Parr, the Holt 45 Caterpillar was small indeed. It had a big heart though and was willing to tackle any job. This one was owned by Walter Fred and Albert Medler. On one occasion Walt pitted it against the gigantic Hart Parr. The complete account is on the next page.

Courtesy, Sherman County Historical Society

BETTING ON A SURE WINNER. Man's will to wager is legend. He'll bet on anything from horses to the date the first snow flies in winter. One of the strangest bets ever drummed up occurred in a summer-fallowed wheat field near Wasco, Oregon on a chill March day in 1908. Here's what happened.

George Hilderbrand's gigantic 17½ ton Hart Parr had broken a ring gear. His problem was how to get the bulky giant out of the field. Pulling it with horses was out of the question. Hilderbrand's thoughts began to focus on his neighbor, Walt Medler, who had a Holt 45 crawler tractor. It hardly seemed possible the little Holt was up to such a tough assignment. Medler was convinced his track-laying iron horse could do the job but neither man knew for sure. They decided to give it a try late one evening. The Holt proved itself by digging in and moving off with the heavy load. That was enough to set the mental gears of the two fun-lovers in motion. The trick they were going to play on the community was already whirring in their minds as they wiped out the tracks of forward progress made by the two gas-powered machines.

The next day word spread, with considerable help from the two instigators, that Medler claimed his little Holt 45 could pull the towering Hart Parr out of the field. The reaction was immediate. Some felt the Holt could do the job while the majority of others were just as quick to react that it couldn't. It wasn't long before people were saying, "'Put your money where your mouth is," and the wager was on. At a designated time the Holt 45 was fired up and managed to handily move off with its ailing neighbor in tow. It made believers of the sizeable crowd that had gathered. No one ever did know if Hilderbrand and Medler fattened their pocketbooks from the stunt. They at least had something to chuckle about the rest of their lives.

DAVID AND GOLIATH. George Hilderbrand walked beside the Holt 45 crawler as it moved off with the Hart Parr. It's obvious he was closely studying the workings of the smaller tractor as he kept pace.

Courtesy, Sherman County Historical Society

WHEAT FIELD PARADE. As the Holt 45 moved out across the summer-fallow plowing with the Hart Parr behind, the crowd fell into step. The picture shows they were walking briskly to stay abreast of the strange parade.

Courtesy, Gordon Hilderbrand

CONVINCED CROWD. By now Medler had made his point. Everyone had seen his Holt crawler tractor do what earlier seemed impossible. It had pulled the 17½ ton Hart Parr and hardly strained a piston in the process. Photographer Raymond grabbed this shot just before the crowd broke up. George Hilderbrand had climbed up to the driver's stand and was ready to steer his Hart Parr in the path of the Holt 45.

Courtesy, Sherman County Historical Society

ON THE WAY HOME. This Hart Parr served in Hilderbrand's wheat fields several years. It later was used to power a sawmill near Hood River, Oregon. Less than 100 of these giant tractors were built by Hart Parr. Their tremendous size spelled their own doom.

Courtesy, Sherman County Historical Society

PLOW POWER. When a Holt 75 crawler grabbed hold of a set of plows there was bound to be some earth turned. This was an Oregon plowing scene in 1918. The 75 weighed 12½ tons.

Courtesy, Al Herman

Chapter IV

Heroes Of The Harvest...

IT SHOULD BE FAIRLY OBVIOUS by now that the wheat harvest was a backbreaking, filthy, dirty job. The men that jigged, sewed, pitched and stacked put up with conditions that would be unthinkable today. They did so mainly because there was no other way.

The population of the small farming communities swelled considerably during the harvest season. By July, each train into town, both freight and passenger, meant dozens of workers piling off and wandering up and down main street waiting for work. They came from everywhere. Some could tell of following the harvest from Texas up through the Dakotas into Canada. Others had farms of their own in the Willamette Valley and were out to make a little extra to help with the mortgage back home.

Each man carried his own bedroll. In some areas it was called a "turkey." It consisted of a blanket or two rolled inside lightweight canvas. The experienced workers could be spotted by their tight rolls. Many of the sack sewers made their talents known by spearing a sack needle through their battered felt hats. Some of their clothes were tattered and worn but they wanted only one thing and that was work.

In addition to wages the men were given room and board but the room wasn't anything to brag about. It was the straw pile. Most outfits worked until eight or nine at night. When the day's work ended the workers didn't hit the hay, they hit the straw.

The 15 to 20 men spread their bedrolls out around the strawpile. They slept on a mattress of straw and radiated out from the straw stack like spokes from a wheel hub. Every head was pointed out and the feet were in toward the pile. The bedrolls were carried in a trap wagon, which doubled as a feed manger for the horses.

Some of the wheat harvest workers traveled so light they didn't even have a bedroll. One such fellow in the Colfax, Washington area was called "Old Joe." Using his coat for a pillow, he literally burrowed himself into the straw pile like a squirrel in a hole.

Other than washing the face and hands, these rugged individuals went to bed with the day's dirt still on them. By the end of the week they went to bed with the week's dirt ground into their skins.

The cookhouse was stationed on a nearby hill. There was almost always some comment by the men as they bedded down about the feminine gender over yonder.

The farmers usually called an early halt to the harvest on Saturdays, so the men could have

supper and head to town for a bath and shave at the barbershop, which was sure to be already jammed with customers.

These men took an old-country pride in their work. In fact, many of the men were from the old country. During idle moments in town or on the farm, talk was largely of the harvest. They reveled in a record day's run. They laughed over a greenhorn's mistakes and told of other harvests in fields far away.

Steam buff Rodney Pitts of Canby, Oregon has written frequent stories for Western Engines magazine that run in the same vein as those told around the straw stack more than a half century ago. In the January, 1965 issue he penned one which proved the old-timers had a sense of humor. It was titled, "Sure Cure for a Sleepy Fireman." The story centered on a threshing outfit in Oregon's Willamette Valley owned by a fellow called Bill.

An early-day tale spinner never launched into a story of the harvest without first pointing at what machinery was involved. Pitts followed the same pattern when at the top of his story he stated that Bill tended his own threshing machine, a 28″ Red River special steel separator powered by a 12 horsepower Advance engine with no canopy, which was fired by wood.

In describing Bill, the author said, "Bill, although a shrewd businessman of German ancestry, was also an incurable clown. He could always be counted on when at the dinner table, in the event the main course ran low, to holler at no one in particular, 'pass to me der fly schwatter—ve is low on der meat running!' or if it happened there was plenty of meat he had still another gambit that never failed to bring loud guffaws from the crew and infuriate the cooks. If someone asked for seconds (and they always did) on the meat, Bill's mouth would pop open and with a feigned look

HARVEST HEROES. There weren't too many smiles in this picture. For the most part the crewmen look a little weary. To the center-right in the front row of men was a fellow who had apparently been wearing goggles. The top half of his face was white and almost clean looking while the bottom half was as black as the inside of a chimney. It resembled the make-up a clown might wear.

These men could wash their hands and face at night in a water basin but there was no place to take a bath for a week at a time. They crawled into their bedrolls with the day's accumulation of dirt ground into their skin. Saturday night, however, everyone headed to town for both a bath and a good time. The local barbershops had bathtubs for rent.

The steam engine barely visible behind the men is a Case 15-45. Its chief identifying feature was the tapered, cast-iron smoke stack. The harvest was in the LaCrosse, Washington area.

Courtesy, Charles Freeburg

of astonishment he would let fly in a voice even the cows could hear, 'Iss that vot that is? Why, I t'ought somevuns iss cott up oldt gum boot!'

"On one particular occasion, Bill's outfit was working away on a warm, late August morning. Firing for Bill this season was lanky 'Norske,' a callow youth of 20 summers who knew barely enough to keep the water showing in the glass. Ray was the tankie.

The regular fireman, Jing, was snoozing with his back propped against the rear wheel of the full tank wagon, a half-round, wood stove type. Mouth open, head lolled forward on chest, his tongue like some limp banner furled and unfurled as his breath passed in and out. The collar of his overalls was pulled back, exposing a generous area of his neck to the warm sun.

"Bill was apt to commute from one end to the other of the outfit for the sake of variety if for no other reason. Everything was going swimmingly at the dusty end; wagons coming in regular rotation; belts tight and the creamy grain pouring from the weigher in pleasing rhythm.

SURROUNDED WITH WORK. It was about 1913 that the harvest crew on the Swannack Brothers farm looked like this. The wheat ranch was close to Lamont, Washington.

Courtesy, T. J. Smith

"But where was Jing? He could see Ray keeping a watchful eye from the smoky end, but that only piqued his curiosity the more. All at once Bill rounded the tank and chanced upon the slumbering Jing who was completely unaware of anything at all—much less a sudden rude awakening. 'Ah,' thinks Bill to himself, 'ashleep yet mit der svitch on iss it? For dis I am baying oudt goot scheckels? Nien!' He cocked his foot for giving Jing a swift kick in the 'piazza'—but how when four feet of snoozling Norseman was shoving the target area into the stubble.

"This called for reconsideration of methods. Clearly some form of retribution was in order, for Bill's German sense of duty and discipline was rubbed raw by the sight he beheld. In his mind the situation simply screamed for some kind of corrective action to be taken. What to do?

"Then his eye caught sight of little Jackie, the property owner's son making his barefoot way through the stubble, slavishly following the broad track left by the outfit and bearing a pail of fresh, cool well water for the now thirsty crew.

"Gesundtheidt! says Bill under his breath—this

is it. A fiendish scheme to teach the sleeper a lesson he'd not soon forget had flashed into Bill's mind at sight of the water boy and he set out at once to put it into action.

"Stepping up to Ray on the engine he pointed to the whistle and held low conversation with him a moment. Then intercepting the child, Bill said, 'So right ober dis vay you should come, Chackie boy, for ve haff here a chentleman very much in need uff your services.'

"Taking a dipperful of the cold water, Bill moved near to the unconscious Jing and held up the grimy index finger of his left hand to alert Ray, then warned the lad: 'So stand chust a liddle furder back, Chackie boy, de dance she iss aboudt to begin commencing.' His sense of the dramatic now at fever pitch, Bill stepped into a position directly over the target; mentally noting to himself how 'soch a dirty kneck should get a varshing vunce between meals!'

"Ray's hand was poised on the baling wire whistle cord that activated the little 2' x 6" buckeye chime which, with the gauge needle at 160 p.s.i. sounded like a shoat caught in the fence.

112

PIPER HARVEST. As near as William Piper can recall it was 1908 when this harvest scene photo was taken on his father's wheat spread near Helix, Oregon. The elder Piper acquired the land in Eastern Oregon under the homestead and timber culture acts. William Piper was standing in about the center of the picture wearing a white shirt and bandana with his arms folded. Brother Oscar was in the second row from the top. He had a mustache and was next to the little boy at the left of the row. The Pipers had hired Jack Leslie and his custom threshing outfit to do the harvest. Leslie was seated in the front position on the horse. There was an average of 37 men and 54 horses performing the harvest. A portion of these men and horses were supplied by the Pipers. The harvest lasted about 35 days. The names of several other harvesters in the group were Ted Norvell, Drapes Sin Clair, Bud Crable and Emil Timmerman.

Courtesy, Piper family

"Bill glanced at Ray to see that all was ready, then his hand dropped, the contents of the dipper started their lethal plunge. At the same time Ray laid back on the baling wire causing the little buckeye to let out a piercing shriek.

"The next sequence is a bit difficult to get down on paper. Suffice it to say that when the chill water smote his neck at the same instant the distress scream of the whistle smote his ear Jing let out a horrible

yell of his own that sure rivaled the little buckeye in volume, though not as pleasing to the ear.

"Leaping to his feet, arms flailing like a windmill in a hurricane, he lurched awkwardly toward the engine while yelping hoarsely, 'Stop 'er! My gawd! Shudderdown!'

"It happened that no other person in the area was in shape to 'shudderdown' at that moment as all were in varying stages of collapse, Bill draped half

WHEAT PROVES KEY TO FUTURE. A threshing crew on the Friedrich Hoefel farm at Ritzville, Washington climbed on top of the sack pile for this shot. In the front row left to right were: second, Emil Hoefel; third, Richard Blum; middle row, first, Fred Hoefel; third, Ed Steffen; fourth, Mr. Knodel; fifth, John F. Hille; sixth, John Hoefel; back row, fifth, Gotthilf Raugust. This was one of many photos David Hoefel published in a private edition book he titled the "Hoefel Family Album." It tells the inspiring story of how his parents with their family of eight children migrated from Russia in 1901 and moved to the wheatlands of Ritzville to seek new opportunity.

Courtesy, A. M. Kendrick

BIG CREW. Three dogs and 40 men handled this harvest. A large operation like this required a small army of men to man the headers, wagons and threshing machinery. This particular crew ran strong to younger men. There wasn't a beard or a really old man in the bunch. The scene was near LaCrosse, Washington.

Courtesy, Charles Freeburg

over the tank wagon, his body wracked by great sobs of delight; Ray holding his splitting sides and gasping for breath. The bundle pitchers just sank into the remainder of their loads and laughed till their throats ached. As soon as the conspirators' wind and strength returned, threshing was resumed with Jing chopping away at the fir knots and Ray back at catching the batflies from beneath the tank team's bellies."

Rodney Pitts not only weaves an interesting yarn but also gives valuable insights to life as it was then. In another story published in Western Engines, February, 1965, he highlighted how a quick thinking engineer averted what could have been a harvest tragedy.

"Kinsey wheat ripened early in the 1921 season around Canby, Oregon. Except for some planting on heavy ground it was all in the shock by July 25. It was the 27th of the month in 90 degree temperature that a sixteen horse Rumley Advance engine was cracking away for all she was worth as the bundle pitchers fed the long straw sheaves into the 33″ x 54″ Russell 'Cyclone' separator. Both men and teams felt the relentless sucking up of every bit of moisture from the skin and mouth. One could barely keep from dehydrating to the point of blowing away.

"The farm owner had, so he thought, solved this problem insofar as the men were concerned by tapping a cask of fine elderberry wine which had been barreled the previous fall. Some partook of this offering between loads and felt benefited by it. The

engineer, Frank, had a small glassful and no more; too much responsibility was in his hands and he recognized this fact very clearly. Among those who took several turns at the glass was Billy. Unfortunately for him his alcohol tolerance was very low. After several trips out for bundles his load seemed to sway and rock extremely far out of plumb. Frank was not oblivious to what was going on. He had been around threshing rigs too many years not to realize cool wine and hot men equals trouble.

"In spite of having to be ever alert for a possible spark that could start a holocaust he concentrated most of his attention on the feeder and the bundle pitchers. There, he reasoned would be the most likely spot for trouble to develop.

"Abner, the outfit owner, was atop the separator, as was his custom keeping an eagle eye out also. A total abstainer, he was greatly perturbed to see the others deaden their reflexes by imbibing and was tense with apprehension lest some awful accident occur. They had not long to wait.

"Billy's turn at the feeder came for the fourth time. His load was not setting well at all; some bundles stuck out too far and made precarious footing. Seizing his fork, after first tying the reins to the rack, Billy made an unsteady lunge at a bundle, missed and jabbed again, this time catching the twine band over the heel of the fork tine.

"Knowing, even in his befuddled state, to keep a firm grip on the fork handle, he tossed and was pulled up by the hung-up bundle, headfirst into the feeder, where whirling band cutters were only inches from his head.

"Back on the smoky end, Frank had just struck a match to light his refilled pipe when the unusual motion caught his eye. His brain registered immediately: 'man in the feeder' and every hair on the back of his neck stood straight out. Before even the sharp cry of alarm from the men at the separator reached his ears, he had, in one sweeping motion, flipped the lighted match into the water barrel and knocked the throttle shut, while his right hand seized the reverse lever and jammed it into the high notch.

"From an even, rhythmic chu-chu-chu the exhaust changed to a reluctant, bumpy paka-paka-paka as the pistons fought against compression and inertia. The entire engine shuddered noisily under the sudden strain. The tortured belt could take just so much.

With a shrill squeal it left the band wheel and went slithering out through the stubble like some huge recoiling serpent, spooking the team on the next wagon up on the belt side so the driver had his hands full for a moment.

"Frightened sober by his sticky predicament, Billy was setting some kind of an unofficial record for the back crawl as he strove to keep clear of the keen edged band cutters. He looked like a frog on a hot skillet.

"Soon, with no power, the steady drone of the separator died away. The speed dropped to a point where the feeder governor cut out when several pairs of hands reached up and helped the badly shaken Billy to the ground.

"There he was in almost as bad shape as moments before because Abner lit into him and administered a severe tongue lashing on being inebriated around

POLISHED UP FOR LUNCH. A little soap and cold water removed the dust and dirt from these men's faces just prior to lunch. Hard work was forgotten when the crew sat down to tables loaded with meat, potatoes and gravy.

Courtesy, Dee Doty

LOOK AT THE CAMERA. Some of the men in this crew were just independent enough they refused to look toward the camera. It appears they purposely turned their backs on the photographer. In contrast, several others looked right down the barrel of the lens.

Courtesy, Sherman County Historical Society

his equipment. 'Why, if you hadda gone in there,' he railed, 'I'da had to of cleaned out the machine,' he said with what amounted to wry humor.

"Then the others took over and by way of relieving tension, commenced joshing poor Billy until he wished his rescue had not been so expertly accomplished. If he could only have been injured just a little. As it was, the 'cure' was worse than the 'ailment.'

"You look whiter'n the nigh horse of the tank team!' said one. 'Yes,' said another, "and I'll bet Billy won't have no need of a bathroom for at least a week, unless he's figgerin on takin a bath!' And still another, 'I don't know if you was mad, Billy, but you was sure hoppin!'

"Belting up again was done quickly and Abner fed Billy's load while he doused himself with cold water to clear his muddled brain. The wine was shunned by all for the rest of the day."

These stories typify a rare breed. The rugged harvest hands didn't need a paid comedian to get a laugh out of life. They found the humor where it was—somewhere between the beginning and end of a hard day's work. They didn't rebel at dirt, dust and grime. Blowing chaff and sticky heat never prompted

anyone to ask for an afternoon shutdown. These men never heard of "blocked nasal passages" but with the clouds of dust billowing out and around the threshing machines they were sure to have had them. The man who couldn't take it never worked in the harvest more than once. Hard work and low pay was their way of life. These soldiers of the wheat fields were the heroes of the harvest.

The small towns of Eastern Oregon and Washington and into Idaho can truly say, "Those were the good old times." They never had it so good and never will again. Those towns and their businesses prospered most when the streets were lined with transient harvesters swapping stories and spinning yarns; when a man's worth was measured by the sacks he sewed and the bushels he threshed; and when poor transportation meant the nearest town was as far as a farmer thought of going for supplies.

Small town prosperity was due for a jolt though as farm machinery manufacturers hit the market with their "combined harvester." The end was in sight for the armies of tough-minded, strong-muscled men who annually answered the call of the waving wheat as it changed from green to gold. This breed of men, rare in quality, soon was to become rare in quantity.

NERVE CENTER. The nerve center of any wheat farm is the home and the equipment sheds. It's here that the plans are made, the books are kept, and the machinery repaired. Almost every activity on the farm funnels through this area.

WHEAT HARVEST REVOLUTIONIZED. The combined harvester opened another chapter in the fast-paced history of mechanical wheat harvesting. The artisic eye of Asahel Curtis caught this dramatic scene in the Big Bend wheat country of Washington August 22, 1915. Powerful mules pulled the combine through the waiting wheat.

Courtesy, Washington State Historical Society

Chapter V

Combined Harvester...

THEY CALLED IT A "COMBINED" HARVESTER because it combined the jobs of cutting and threshing into one operation—into one machine. In a single sweep the grain was cut, separated and sacked. There was no need to haul the headed or bundled wheat to the separator. Instead of bringing the wheat to the thresher the combine took its thresher to the wheat and did the job while moving through a field.

The 1929 issue of Combine yearbook laid out the history of the combined harvester in considerable detail. It pointed out, "the combine in America appears to have gotten its real start in California. No doubt due to the fact that climatic conditions in that state were particularly favorable to successful combine operation, and wheat growing was a more or less prominent industry. As early as 1867 we find that D. C. Matteson built a combine at Stockton, and that patents were granted to several residents of the states of California and Oregon in the early '60s.

"A few years later, in the early '70s, several men on the Pacific Coast were apparently alive to the possibilities of the combine and began to incorporate their ideas into actual machines.

"By 1887 several machines were scattered over the state of California and by 1890 several concerns, Best, Holt, Houser and Haines, Mingee Shippee, Young and Berry, were in production and ready to meet the requirements of the Pacific Coast wheat growers.

"These early machines differed little in principle from the combines in actual use today. Wood construction, which was vogue at that time, was used. This material made the machines very cumbersome which was further augmented by the fact that the idea of the early combine manufacturer was to get as much capacity as possible out a single unit.

"Machines cutting a swath 30 feet wide or even wider were constructed and as their weight ran up to 15 tons or more it required as many as 40 horses or mules to pull them, especially where the land was rolling.

"To meet the hilly conditions of Washington it became necessary to develop a leveling device which would compensate for the unevenness of the fields. This gave rise to what is now known as the 'hillside' as against the 'prairie' type of combine. The hillside type of combine is still operated largely by horses because of the fact that it is very difficult to operate tractors on such steep grades."

The first machines, wood-framed and heavy, didn't explode onto the harvest scene as overnight sensations. In those days farmers were somewhat slow to change but the machines did make a steady progression at winning converts. In Adams County Washington, for instance, the record shows

32 HORSEPOWER. They called it the "combined harvester." The reason was simple. It combined the jobs previously done by both the header and the threshing machine. In one operation, the huge harvester cut the wheat and separated it while moving through the field. It cut down on manpower requirements dramatically, but at least 32 horses were still needed.

The harvest picture on Charles Pool's ranch near Moro, Oregon was typical of how the horses were hooked to the combine. Five rows of six horses each comprised the bulk of the animals. Two horses out front were the leaders. They were chosen for their speed and intelligence. Some farmers figured the job of turning all the other animals around corners was too much for just two horses so they placed three in the lead for a total of 33. Amazingly, the driver controlled this massive team with just two reins running out to the leaders. In rare instances when conditions were particularly hilly the number of horses used ran into the 40's.

The switch from stationary threshing machines to combined harvesters had a tremendous impact on the labor force needed during the summer wheat harvest. Thousands of transient "bindle-stiffs" who for years had made the wheat harvest possible found themselves no longer needed as machines began doing more and more jobs.

Courtesy, Sherman County Historical Society

that the first ground-powered combine was used on the John Gillett farm 14 miles northwest of Ritzville in 1898. Farmers came from 60 and 70 miles around to watch it. This wasn't the first combined harvester in the northwest but it was among the early ones.

Before the machines had proven themselves, it was always the progressive, innovating farmers in a community who were the first to plunk down the cash for one of the "new fangled gadgets." There was one particular old wives' tale which held many back from making the switch. Scores of wheatmen still clung to the idea that cutting the grain and separating immediately was no good—that it had to go through a "sweat" in the stack before it was ready for milling.

By 1906 the machines were catching on and 13 were introduced in the Washtucna, Washington area. As more began appearing in the wheatfields they became known simply as "combines" rather than "combined harvesters." The transition from headers and threshers to combines stretched itself out from 1900 to 1920.

When a farmer went from a stationary operation to the mobile combine his harvesting crew suddenly plummeted from an average of 20 men to just five or six. It's obvious then that the demand for harvest hands began tapering off as more and more farmers bought the new machines.

The first combines were ground-powered rigs. Power, in other words, to run the sickle bar, header reel, belts, pulleys and threshing mechanism came from one of the combine wheels called the ground wheel. As the combined harvester moved forward, power was transferred to the working parts by chains and sprockets.

32 HORSEPOWER

The combine required 32 horsepower to function. This was the most amazing aspect of the entire operation. This horsepower requirement referred to the real, live, four-footed, hay-burning variety of horsepower, not the theoretical horsepower rating given to the drawbar and pulley strength of a steam engine.

Imagine the spectacle—32 horses in one hitch

pulling a machine through the fields which gobbled up wheat in one end and discharged filled sacks out the other end. There were times in tough-pulling country when they used 44, even 50 horses or mules to do the job.

The usual team of 32 had two horses in the lead followed by five rows of six animals each. On curves and corners it was up to the leaders to swing the rest of the harnessed workers around. Some teamsters felt that was too big a job for two animals so they used three in the front line.

To the onlooker, the driver had the most glamorous job on the combine. He was perched high on the end of a ladder which slanted out over the horses. It was like riding in the prow of a ship. The hills were the waves.

There were times just before reaching the peak of a hill that the driver was projected so far up and at a backward slant that he couldn't see the horses in front of him. They were already over the hill and out of sight. Even the most experienced drivers had a feeling of relief as the combine pulled over the hill and the horses came back into view.

The comparison was the same at the bottom of a hill. It was the trough of the wave. With the horses already starting up the next rise and the combine coming down the last, the driver was suddenly thrust right down in and among the draft animals just a few feet above the ground.

Although it seldom happened the combine crew knew only too well the danger of straw piling up under the bull wheel on the down side of a hill. It occurred just enough that the very thought sent shudders down their backbones.

Big and heavy as it was, the drive wheel could actually lose contact with the ground if enough straw built up. As it wedged in under the wheel, lifting it up, the mass of straw became a runner, putting the entire combine on skids. An overturned combine or pile-up of some sort was inevitable.

For the driver, such a runaway accident was scarier than any roller coaster ride and considerably more dangerous. As the combine began careening down the hill the teamster was no doubt wondering why he hadn't chosen a more genteel occupation.

It was the wheeler horses, the ones closest to the

GROUND POWER. This unusually fine shot gives a clear view of the ground wheel and how it was chained to the rest of the machinery. One of the decided disadvantages of this type machine was that it worked only when the combined harvester was moving forward. The minute the forward progress stopped, so did the threshing mechanism. The driver on this machine was not perched out and over the horses on a ladder as with the majority of combines.

Courtesy, Charles Freeburg

combine, that faced the greatest danger though. Frequently two or three would be crushed in the terrifying downhill flight. Fortunately, such accidents were rare.

Driving 32 horses or mules was an art in itself. Incomprehensible as it may seem, the driver guided his charges with just two lines. If he had a jerk-line mule setting the pace he needed only one line.

The normal procedure was to have a line running to each of the two lead animals. The fastest, smartest animals were always up front. When the driver tugged a rein indicating a turn the leaders had to swing all the horses in the desired direction.

The jerk-line mule was even more of a phenomenon. Paired with another mule he was the left-hand leader. To go to the left the driver gave a steady pull on the single line. As the mule went left so did all the other mules. A series of short jerks caused the lead mule to throw his head up. This in turn activated a G-string to pull his head to the right. The lead mule then turned right followed by the rest of the team.

A G-string wasn't needed when the mule was trained to the signals. He quickly responded to the left when he felt a steady pull on the rein and turned right when there were a couple of jerks.

They used to say a driver had to be at least as smart as the horses he drove or he wouldn't last long. That may have been true. Equine I. Q. rated high among those animals that had a lazy streak and looked for ways to get out of work. Usually they followed the simple ruse of lagging back in the traces and merely walked along instead of pulling. Some got so clever they could keep the traces taut and look like they were pulling when they weren't at all.

Most drivers didn't use a whip to remind a wily horse that he had a load to pull. When mule skinner's language failed, the driver resorted to rocks or clods. They were kept in a small built-in box next to the driver. The offender snapped to, at least for a little while, when he felt a sting of a pebble bouncing off his rump. Daydreams of throwing baseballs instead of rocks inspired many drivers into becoming excellent marksmen. Other skinners didn't waste their time with rocks to gain an animal's attention. Instead, they used air guns.

Ted Porter of Walla Walla started as a mule driver in 1923 when he was 15. By then the combines had motors on them to run the machinery. The work animals still pulled the combine forward but fewer were needed since the harvesters no longer relied on ground power to activate the mechanical apparatus. The number of horses dropped to 27 on most rigs with the addition of a motor, but this depended on the topography and the size of the machine.

THE PALOUSE HILLS. These Palouse hills proved to be some of the most productive wheat lands in the world. They also proved to be the most difficult to harvest. The huge, cumbersome early-day tractors didn't have a chance on these steep slopes. The first steam and gas engines were so big and heavy that it would have taken most of their power to just get themselves up the hills let alone pulling a heavy combine behind. Even if they could have powered themselves around the steep inclines, chances are the gentle giants would have toppled over from top-heaviness and poor balance. Horses were the only answer until crawler tractors and the smaller, yet powerful wheel tractors were developed.

Fewer mules were pulling this combine. Twenty-seven could do the job since a motor was added to later model harvesters. The motor powered the machinery while the animals pulled the combine through the fields.

Courtesy, Allis Chalmers Co.

123

SIDE-HILL LEVELER. One of the biggest steps forward for the combined harvest was a sidehill leveling device introduced by the Holt Co. in 1891. It was a particularly important development for the harvest of the hilly Northwest. This picture shows how it worked. When the photo was snapped the combine was on a slope. The header slanted up the hill and yet the main body of the combine was level. Without the leveling device the wheat going through the machine would accumulate on the bottom side of the sloping cylinder in a bunched-up slug. As a result, it was impossible to do the most efficient job of threshing. The leveler eliminated this problem.

The combine man was responsible for seeing that the machine was kept level and also that all the machinery was functioning smoothly. He was the man at the left of the picture.

The same basic leveler is used on modern-day combines as shown in the picture below.

Courtesy, Wayne Doty

MAKING THE TURN. It was no simple matter getting 33 horses and a combine around a sharp turn. These three pictures taken from the 1918 annual report of M. S. Shrock, County Agent from Umatilla County, Oregon depict an old-style ground-power machine, drawn by 33 horses, in the act of making a sharp turn while finishing a three cornered piece. The second photo portrays the three lead animals swinging the rest of horses in an arc. The last shot shows the turn completed. Note the bulk grain wagon drawn by four horses taking the grain as it comes from the combine.

Courtesy, Oregon State University Archives

Now a store owner, Porter vividly remembers his early experiences. When asked which was best for wheat work, horses or mules, Porter sided with mules. He had a string of arguments in their favor.

According to Porter, mules were far and away the best performers. For one thing he said, "Mules could stand the heat better. You couldn't heat-founder a mule like you could a horse. When a mule had a combination of too much heat and work he stopped. You've heard of a balky mule. That's just what he did. He balked and heaven or hell wouldn't budge him. That could be exasperating but I always figured

it was better than a horse working until it dropped dead in its tracks."

At this point, Porter was just warming up in his stand for the mule. He also pointed out that mules were less nervous than horses; they never went into fences and horses did; and mules were supposed to have had the good common sense not to step into badger holes while horses would. Interestingly enough, though, the county censuses taken during that era showed horses easily outnumbered mules.

Ted Porter speaks almost affectionately about his days as a mule skinner on a combine but adds, "The

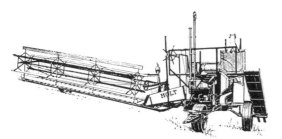

Holt Model 36, Hillside Type.

Wood Bros. New Model, Prairie Type.

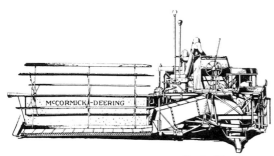

McCormick-Deering No. 7, Hillside Type.

Avery Model B, Prairie Type.

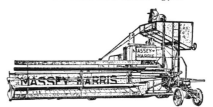

Massey-Harris 9A, Prairie Type.

Case Model W, Hillside Type.

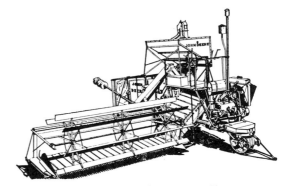

John Deere, Model 1, Prairie Type.

COMBINE GALLERY. A few of the pictures run in the 1929 Combine Year Book of harvesters on the market at that time.

Courtesy, F. Hal Higgins, University of California at Davis

Gleaner Model R-12, Level Land Type.

PHOTOGRAPHER'S FAMILY. On a hot day in August of 1903, Mr. Gaffney and Mr. Carey, the owners of the above combine, hired commercial photographer Tom Richardson to shoot this picture. Their ranch was located six miles out of Sprague, Washington. Richardson brought along Callie Epley, his brother Milt Richardson, and Milt's wife Etta and their son, Jeff. This was one of the photos taken which included not only the combine crew but also the Richardson entourage. Milt Richardson is on the upper level of the harvest machine at the right wearing the dark suit. The women can also be seen topside. Little Jeff couldn't have been in a loftier, prouder position as he shared the driver's seat. Years later, Jeff Richardson was active in the Steam Fiends Association, an organization dedicated to keeping the memory of steam alive. The combine appears to be a Holt. It cut 20 feet of wheat and was pulled by 32 horses.

Courtesy, Jeff Richardson

job wasn't a picnic. It wasn't quite as glamorous as it looked. For one thing, it was extremely dusty. All the mules below and ahead were churning up dust. You were trapped in it. It absolutely encased you. If you had a tail wind you not only got the dust of the mules but also the chaff from the combine."

He described two things which occasionally happened that caused his heart to skip a beat or two. "Say the outfit was on a sidehill. That had you a little nervous to start with," he warned. "Then let the wheel on the lower side drop into a badger hole and you thought she was going over for sure.

"It was just as scary when we'd be going around a steep hill and we'd hit soft dirt and the whole rig would slide sideways. Don't think that wouldn't shake you up a little bit," Porter recalled.

The former mule handler went on to say that the steady pull of the reins all day left his arms aching the entire night. "It wore your arms out because once you picked up the reins it was a steady pull from start to finish."

In outlining the activities of a normal day, Porter commented, "You were always up by 4 a.m. to get the mules ready. The driver was responsible for nine head. The rest were cared for by the sack-sewer, sack-jig and header-tender who took six apiece. We fed, curried and harnessed them."

He emphasized that currying the stock not only

PHOTOGRAPHER A. M. KENDRICK STARTED ON A COMBINE. A. M. or "Bert" Kendrick spent his early childhood on the wheat ranch of his father A. T. Kendrick. It was located near Cunningham, Washington. In 1904, this photo was taken of the family and crew during the harvest. Young Bert was on the upper platform with the header-tender. Years later, Kendrick became a commercial cameraman and took hundreds of harvest scene pictures himself. Others on the combine were Charles Howes, driver; Mary Kendrick, Bert's sister sitting on the highest level in the black dress; Anna Hale, a neighbor; John Kendrick, little boy on sewing platform; A. T. Kendrick wearing a white shirt at the foot of the ladder; and "Old Tom" the combine man leaning on the upper railing who was "badly bent up with arthritis but good," according to A. M. Kendrick. The rest of the crew was from Missouri. "Missourians and mules just went together," says Kendrick. He related that as a boy he used to spend every spare moment at the bunk house listening to them tell tales of outlaws they knew ranging from the Youngers and the James boys to other desparadoes of the time. Kendrick remembmers he listened in total fascination.

Courtesy, A. Kendrick

made them look better but it could lead to trouble if you didn't. If a mule or horse wasn't curried down the dirt could build up in the hair. Sores often resulted as the harness rubbed the accumulated dirt.

Porter continued his description of the early morning chores by saying, "Once we fed the mules their wheat hay and curried them, we harnessed them, had breakfast, hooked up and were in the field by 6 a.m. We worked until 11:00 and started again at 1:00 and continued at least until 6:00.

"In those two hours at lunch we weren't under a shade tree resting, I can guarantee you that. The mules had to be brought in and fed. Feed racks were always pulled to wherever the outfit was working. It was the driver's job to work out any sores the mules might get from the hard work. He normally rubbed salve on the sore spots to speed up the healing. The sores usually came in the shoulders where the pressure of pulling was the greatest. One of the best preventions," he said, "was in making sure each mule had a collar that was big enough for him."

"At night we tied the mules to the feed rack. It was the same process only in reverse of taking the harness off, feeding the mules and giving them a curry job. The harness was set on the ground behind each animal. We had a special way of setting the harness so it was easy to grab the next morning. The hames and collar were upright as they leaned against each other and the rest of the harness trailed back. We used this arrangement when we were way out in a big field. There were many times though when we were close to a barn and could put the harnesses inside overnight."

Mr. Porter emphasized that the mules were never kept in a barn at night during the summertime harvest for two reasons. For one thing it was cooler and more comfortable for the animals outside. Secondly, the barns were almost always filled with hay. There was the chance this hay could go up in flames from spontaneous combustion. The farmers didn't want to take that gamble.

The work animals wintered outside also, getting by on the food they gleaned from the stubble fields. When the weather got really tough, the farmers brought their four-legged charges in for feeding and shelter.

During harvest the driver had the company of four other workers who rode the combine. They were the header-tender, combine man, sack-sewer, and sack-jig. There were also a couple of men picking up sacks in the field.

The combine man was usually the foreman of the crew. He rode on top of the combine just over the two rear wheels on the right-hand side. Like the separator man he kept an elert eye on the machine's smooth operation.

It was also up to him to keep the combine level as the outfit snaked its way around the hills. This he could do because of a sidehill leveling device introduced by the Holt Co. in 1891.

MULE-SKINNER NEEDED NERVES OF STEEL. These so-called gentle, rolling hills provided many anxious moments for mule-skinners. There were times when coming up over the top of a hill that the team was already down the other side out of sight. That tended to step up the heartbeat of even the most experienced drivers. Another agonizing experience occurred when the downhill wheel of the combine dropped into a badger hole. Not only the driver, but the entire crew thought for sure the whole outfit was going over—sometimes it did.

Courtesy, Ed Lawrence

COMBINE ON HER SIDE. If a hillside combine was going to tip over, logic would dictate that the machine would fall downhill instead of up. The combine in these pictures hadn't taken a course in logic, however. It was just contrary enough to want to tip uphill and that's just what it did. According to Elick Huff, sack-sewer on the combine, an extremely sharp turn at the bottom of a small gulley caused the mishap. Huff is shown on top of the downed combine waving his hat at the cameraman. He said the accident could have been worse considering the machine and crew were back to work late the next day. The combined harvester was put back on both wheels by hooking the horses to it and pulling the crippled implement upright. Huff added, "There was considerable fixing up to do and parts to be replaced but once she was fixed she ran just as good as new—well, almost anyway."

Elick Huff lived in Walla Walla, Washington and was a painter by trade. When wheat harvest rolled around, he traded his paint brush for a sack-sewer's needle. He said this accident was about the scariest experience he had in his years of harvesting. The accident occurred on the Ben Cole ranch near Walla Walla in 1931.

Courtesy, Elick Huff

UPSY-DAISY. The Palouse hills are just as steep today as they were at the turn of the century. It's just as difficult to hold a combine on them as it used to be. Every year several of the harvesters lose their grip and virtually destroy themselves before coming to rest at the bottom of the hill. This was one modern combine that harvested no more crops. Jake Ottmar of Colfax, Washington was surveying the damage of a mid 60's pileup.

Courtesy, Colfax Gazette

Hal Higgins, foremost agricultural historian with the University of California at Davis, says that only after this ingenious device came into use could the combine be really practical in the extremely hilly Palouse region of Southeastern Washington and Northeastern Oregon.

Without the leveler a combine leaning over a hill did a very inefficient job of threshing. The wheat jiggled down to the bottom side of the threshing cylinder and piled up there. It came through the concaves in an impenetrable slug. This resulted in most of the wheat kernels not being separated from the stalk and so, passing out the back end with the straw.

The leveler corrected all this. Those seeing the leveler working for the first time stood fascinated at the effectiveness of the device in operation. They stared as they saw a combine tilted over on its side as it followed the slope of a hill and yet the machine's threshing cylinder remained level.

The wheat passed through evenly with the kernels going into sacks instead of out the back end onto the ground. The basic principle of this early leveler is still used on today's combines.

One added advantage the leveler gave the combine was an adjusted center of gravity. As a result the big machine wasn't so tempted to topple from the slope. Even with the leveler, the almost perpendicular sides of some wheat hills took their toll as several combines turned over each year.

Hal Higgins conjectures that thousands more acres of hilly northwest land were more adaptable to wheat after Holt came out with the leveler.

The combine man greased the machinery at noon and at night. Frequently he was assisted in this job by a grease monkey who came out from the farmstead.

The sacking platform was usually covered with an improvised canvas shade roof. It was commonly referred to as the "doghouse."

The sack-jig's job at the threshing machine and on the combine was the same. He put sacks on one spout, took them off the other, jounced them, and set them in front of the sewer. The sack-sewer sat next to the chute. As he finished his sewing he gave the sack a flip and it fell down the chute.

Originally, the sacking platform sat low on the combine, just a few feet above the ground. The short chute could hold only three sacks. Later developments raised the platform up near the top of the harvester. The chute was then longer and held five sacks. When the chute was full the sacks were dumped in as near a level spot as could be found, minimizing the trouble the pickup team had in loading the sacks from the ground.

An incident involving a sack-jig and sack-sewer appeared in "The Palouse Story" prepared by a history committee of interested Palouse citizens in 1962. It told how at the crest of a hill the hitch on the

SACK DUMP. Some of the earliest combined-harvesters didn't have a sack chute. This picture shows how the sack sewers piled the sacks on a small extended platform. They dumped piles of four or five at a time on the ground.

Courtesy, Francis A. Wood

combine broke, pulling the driver from his seat as the horses went suddenly and unexpectedly forward. It went on to describe the incident by saying, "Backward down the hill rolled the heavy machine, with the boss' saddle horse tied behind, zigzagging to keep out of its path. The sack-sewer jumped down from the high-up doghouse and the jig followed. Catching up with the horses that had become grounded, it was necessary to cut harness to unscramble the mess.

"At the moment it wasn't funny, but afterwards the sack-sewer used to say, 'And Ralph, that jig, when he jumped, couldn't find any other place to light in that whole forty-acre field but square on top of my back'."

ON THE MOVE. A fine action shot of horses and men engrossed in the harvest. Four of the six hands on the combined harvester were totally oblivious to the camera. One of the men at the rear end of the machine was the combine man. The other was just along for the ride.

Courtesy, Sherman County Historical Society

BABY HOLT. The designers of the first combined harvesters were out to make the machines as big as possible. Hill-side combines frequently had 24 foot headers. In the plains states, combines cut up to 34 foot swathes. The theory was that bigger machines could harvest more wheat. That was fine for the large farms, but smaller landholders began to raise their voices for a less expensive model. Combine manufacturers responded. This was a baby Holt which came out about 1910. It cut a 12 foot swath, was pulled by 18 horses and featured a crew of four men instead of the normal five. The scaled-down Holt had one man doing the jigging and sewing of sacks rather than two. The above machine was owned and operated for two seasons by Omer J. Sayers of Moro, Oregon. Models requiring even less manpower hit the market within a few years.

Courtesy, Sherman County Historical Society

ADAPTATION. This machine isn't what it started out to be. When Ben Oestreich originally bought the machine it was a ground-powered combine. Then Ben got busy and added a motor, cutting down on the number of horses needed to pull it. He eliminated sacking by adding a bulk tank on top. This was the Oestreich farm as it looked in 1936, near Ritzville, Washington.

Courtesy, A. M. Kendrick

ONLY TWO MEN NEEDED. McCormick and Deering really set off a bombshell when they announced in 1917 they were coming out with baby harvesters that were drastically reduced in size and manpower requirements. On the market the next year, most of these baby combines had a nine foot cut, were ground-powered, could be pulled by eight to twelve horses, and required only a two-man crew. Since they weren't equipped with a leveling device and had no motor they couldn't be considered extremely efficient. They had one big advantage to the small farmer and that was their price. It ran in the neighborhood of $1600. In many areas they sold so fast dealers found themselves virtually working around the clock making deliveries. Curtis made this shot in about 1922 on the H. D. Irvin ranch near Rathdrum, Idaho.

Courtesy, Washington State Historical Society

BULKING WHEAT. No sacks were needed on this combine since the grain was bulked. It was still practical to use horses or mules to pull combined harvesters as late as 1927. In fact, the Wheat Growers Economic Conference of 1926 which met in Moro, Oregon reported that on farms of less than 1000 acres wheat could be produced at less cost per acre and per bushel with horses alone than with tractors and horses. Carroll Sayrs seemed to follow that line of reasoning as he went about his harvest with this outfit in 1927. His motor-powered Case combine was still horse-drawn and this was the story on many farms. The wheat conference also brought out that motor-driven combines such as the one pictured harvested wheat at less cost than ground-powered combines. Ground-powered machines had all but faded from the scene by this time.

Courtesy, Sherman County Historical Society

A BREATHTAKING SIGHT. This harvest scene encompasses two rivers, three counties, and two states. The rivers are the Deschutes in the foreground and the Columbia in the distance. The counties include Wasco and Sherman in Oregon and Klickitat in Washington. The states, of course, are Oregon on the near side and Washington on the far side of the Columbia.

The combined harvester belonged to George Wagonblast of Dufur, Oregon. Strangely, Wagonblast finds fault with the beautiful panorama shot. He sees only one thing in this picture and it is not the vast scope of the country-side. It is rather the one white mule among all the black mules. According to the elderly Wagonblast, "that white

mule was in the lineup only a short time but naturally, that's when the photographer had to come out." The shot was taken in 1931 but George Wagonblast was still chafing about the white mule into the late '60's.

The Wagonblasts went from a stationary thresher operation to horses and a combined harvester in 1923. They switched to mules in 1929 since the horses were so susceptible to mountain fever. The mules were purchased for an average of $175 apiece and sold in 1942 for a $45 average.

George Wagonblast was the combine man. He was on the left in the photo. The driver, Stanley Huston; header-tender (behind the wheel), Carl McCurdy; sack-jig, Edwin Wagonblast (standing); sack sewer, John Stark.

Courtesy, Mel Olmstead

C. O. CAMP, PAMPA, WASH. 1913.
PHOTO BY HUTCHISON ENDICOTT, WASH.

WOMAN ABOARD. Community photographers traveled from field to field during the harvest taking pictures of every outfit that requested it. Farmers that failed to have a photo taken were in the definite minority. The photographs featured not only the harvest crew but invariably the women folk and often the children climbed aboard the combine to be a part of the picture gallery.

Courtesy, Sherman County Historical Society

CAMP MEETING. Brothers Oscar and Cleaon Camp played a vital part in their harvest of 1913. Oscar was the roustabout who tended to the odd jobs. He was at the left on the horse. Next to him on top of the combine was Cleaon, combine man. Cleaon later became a popular hardware and implement dealer in LaCrosse, Washington.

Courtesy, C. O. Camp

ALL EARS. The ground-powered combined harvester of G. E. Webb, Lind, Washington was pulled by 27 big-eared mules. The date was August 24, 1916.

Courtesy, A. M. Kendrick

PUBLISHED BY
IMPLEMENT TRADE JOURNAL CO.
INCORPORATED

Graphic Arts Bldg., 10th and Wyandotte Sts.
Kansas City, Mo.

OFFICERS AND DIRECTORS
CLIFFORD F. HALL, President
CARL J. SIMPSON, Vice-Pres. and Treas.
CARL W. HERTEL, Secretary

THE WEEKLY
IMPLEMENT TRADE
JOURNAL

Founded 1886   Issued Saturdays

SEMI-MONTHLY HARDWARE DEPARTMENT
An Illustrated Section Devoted to Hardware and Kindred Interests
Appears Twice a Month in This Paper

BRANCH OFFICE, OMAHA, NEB.
1019-20 Woodman Bldg.
W. J. ROSEBERRY, Manager

FRED MILBURN, Western Representative
4408 Magnolia Avenue
Chicago, Ill.

Entered at the Kansas City, Mo., Post Office as
Mail Matter of the Second Class

SUBSCRIPTION RATES:
United States and Mexico - - - $2.00
Canada and Foreign Countries - - 3.00

A Baby Combined Harvester Is Born

QUESNELL? Yes. C. Quesnell, that's his name. But there isn't anything else queer about him, for the wheat-growing section of the Pacific Northwest takes him and his inventions mighty seriously. Mr. Quesnell is the inventor of the one-man combined harvester. He has invented other harvesters, too, but this latest one is bidding for the attention of the small-farm wheat-grower of the "Inland Empire" just now.

Some years ago Mr. Quesnell brought out a combined harvester which required eight

the grain is carried by a draper directly back from the cutter-bar to the cylinder, which is six feet wide. There it is threshed and deposited on a roller sieve, which separates and cleans the grain. The straw goes on over and out the back of the machine on the ground. The tailings are elevated back over the draper and through the machine.

Approximately twenty bushels are held in the hopper. The pause for emptying the threshed grain into sacks or other receptacles

by the harvester companies. To these machines the grain-growers owe much of their success. And they will probably be sold in large numbers for years to come.

But just now the agricultural observers of that part of the country, so far as they are concerned with the harvesting of small grain, are watching the development of this type of small combined harvester which has met with so much encouragement and which is new to the country. Also they are observing with keen

THE NEW QUESNELL BABY COMBINED HARVESTER AND THRESHER

horses and cut nine feet. Now he has taken a long step forward by developing his one-man outfit, which requires but four horses and cuts 7½ feet.

Farmers of the Middle West know very little about the combined harvester. This remarkable machine cuts and threshes small grain at the same operation. Of course, the straw is left in the field. Some outfits are pulled by horses and others by tractors.

In his one-man outfit Mr. Quesnell has endeavored to emancipate the wheat farmers of the Pacific Northwest from the help problem, and from the delays and expense of the threshing season. The builders of the new Quesnell combined harvester assert that it makes it possible for the farmer to reap what he sows with he same horse power that put in his crop.

The machine operates much like a header. The horses are hitched behind it. As it is cut

affords the horses time to rest, that is, if it isn't pushed by a tractor. Unless a gas engine is used as auxiliary power to operate the cutting mechanism, six horses are required.

One man, who drives the horses and works the stand as with a header, cares for all the machinery. There are but three levers to manipulate. One raises and lowers the platform. One levels the cutter-bar independent of the position of the platform. The third adjusts the sieves.

Combined harvesters are very much the rule in the "Inland Empire" among the big wheat ranchers. The smaller grain farmers use binders and have their wheat threshed from the stack or shock just as is done in the Middle West. It is the hope of Mr. Quesnell that with his new machine the smaller farmers will take advantage of their opportunity to save labor.

Thousands of binders, of course, are sold to the Pacific-Northwestern farmers every year

interest the application of more and more mechanical horsepower to the operation of the larger combined harvesters. With the former the farmer is enabled to reduce the number of horses on his farm as well as the amount of his help.

With the latter, where he has the capital and the acreage, he is enabled to harvest an enormous volume of grain at a comparatively small expense. For the Easterner the sight of one of these huge machines cruising across a wheat ranch, cutting and threshing the grain as it goes, is one never to be forgotten. But this latest Quesnell machine, of course, is built on a smaller scale.

As the war continues the English pound sterling, so-called, goes down in value and the American dollar goes up abroad. At the present rate the dollar will soon be well known from one end of the world to the other.

NORTHWEST COMBINE MANUFACTURER. An article in the Weekly Implement Trade Journal pointed out the genius of Cornelius Quesnell.

Courtesy, F. Hal Higgins, University of California at Davis

QUESNELL COMBINE. This is another of many combined harvesters designed by Cornelius Quesnell. The rig shown was both pushed and pulled. There were four horses up front and eight behind. One of Quesnell's chief goals was to cut down on manpower requirements. Only three men were required to operate this machine. The baby combine included an eight foot cut and an eight foot cylinder. Quesnell and A. M. Anderson patented their first models at Moscow, Idaho and later developed improvements at the Multnomah Iron Works in Portland, Oregon.

Courtesy, Ted Worrall

LIKE DRIVING A HEADER. This was apparently a model of Quesnell's pusher-type combine. Inspiration for its design no doubt came from the earlier header implement. It had the same three-wheel arrangement. The only difference was that seven or eight instead of six horses were in the rear pushing. Most of the steering was done with the back wheel. The teamster controlled this single wheel by a stick running between his knees as with the header. The driver of one of these rigs didn't have much time to himself. He not only had to keep his eyes and ears tuned to hear mechanical problems, handle the horses, and steer the machine but it was also his responsibility to raise and lower the cutter head. This varied with the height of the wheat and the unevenness of the ground. The long lever extending back made it possible for the driver to make adjustments from where he stood. The sack-sewer worked on the far side of the machine.

Courtesy, Sherman County Historical Society

WHEAT HARVEST RARITY. It was 1910 that George Drumheller had his five combined harvesters in the same field. Each machine was pulled by 33 mules for a total of 165 animals in one operation. This shot had great popularity as a post card picture at the time. The combines were called "Oregon Specials" and were made by the Holt Manufacturing Co. They were ground-powered. Dead hitches were used resulting in the harvesters being almost animal killers. For some of his lead mules Drumheller was reported to have paid as much as $600 each. It was also reported as fact that the wheat crop harvested by the five combines that year brought Drumheller a check of $125,000. The Drumheller ranch was located in Walla Walla county, Washington on Dry Gulch creek.

Courtesy, F. Hal Higgins, University of California at Davis

WHEAT FIELD CLOWNS. W. A. Raymond was an unrecognized artist in his community of Moro, Oregon. A master with a camera, he also had a sense of humor. In this instance he caught the W. A. Woods crew as they were horsing around during the picture-taking break. Most community cameramen of that time would have taken only the more serious, posed type of shot, but not Raymond. He saw the humor of the situation and snapped it. It's said that Raymond often trekked back into the mountains for several weeks at a time just to study and photograph cloud formations and patterns in nature. To the people of Moro he was known as an all-around handyman, good at carpentry or cement work and a cameraman on the side. Upon his death many of his pictures and negatives were destroyed before the value of them was realized. These early-day photographers did an amazing job with the tools they had.

Courtesy, Sherman County Historical Society

BULKING. According to Umatilla County Agent M. S. Shrock in his 1918 annual report this was "One of the modern Oregon Special combines, cutting 20 feet with a gas engine mounted on the combine for power and drawn by a 75 horsepower Holt caterpillar. The bulk grain wagon is 'picked up' without stopping and drawn until filled, then 'dropped' and another one is picked up."

The owner must have had a patriotic bent considering the flag flying from both pieces of equipment. The way the flags were standing straight out indicates there was a fairly stiff breeze blowing the day the shot was taken.

Courtesy, Oregon State University Archives

WAGON HOOKUP. The device to which the bulk wagon was attached was quite ingenious. It eliminated the need for horses or mules to pull the grain wagon beside the combine as it made its rounds.

Courtesy, Oregon State University Archives

GAS POWER REPLACES HORSE POWER. The photos of Asahel Curtis covered a wide span of time—from the early days of horses and steam to the mechanized harvest. Covering the state with his camera in 1939 he caught this scene of Casper and Joe Flaig harvesting near Spokane, Washington. A Cat tractor was pulling the combine up a fairly steep grade. The date was August 9.

Courtesy, Washington State Historical Society

CRAWLER AND COMBINE. It was easy to get in the doghouse when working on a combined-harvester. The doghouse was not a state of mind but rather a platform where the sacks were sewn. More often than not the doghouse was walled on the sides and roofed on the top by worn out canvas drapes. This combine is an example of how the workers fashioned the semblance of a shelter from canvas drapes and scrap lumber. A Best 60 tractor pulled the combine.

Courtesy, Sherman County Historical Society

EXPERIMENTAL MODEL. Leading Washington wheat rancher, Paul S. Hofer, had a hand in the design of the first steel combines. In making the shift from wood to steel the Holt Co. invited Hofer down to their Stockton, California factories as a consultant in 1929. Based on their experience in the wheat fields, he and several others fed ideas and suggestions to the design engineers. One of the first models was shipped to Hofer's Top Wave Farms in Prescott, Washington in 1930 to be used under field conditions. At the end of that year the harvester was returned to Stockton for improvements. The updated version was back in the Hofer wheat fields as the next harvest began. The Holt Co. left it there for the next 15 years to observe its performance.

The RD7 Caterpillar tractor pulling the experimental combine was also on trial. It was one of the first crawlers in a frame of its own. Placed on the ranch by the Caterpillar Co. it is there today and still in use.

Hofer says the family farm got its name from his mother. On first seeing the country, Mrs. Hofer said the rolling hills reminded her of the ocean waves. The result was the name, Top Wave Farms. *Courtesy, Paul S. Hofer*

A PAIR OF HOLTS. Two Holts team up in this harvest scene. A Holt 75 tractor pulls a motor-driven Holt combine in Eastern Oregon. The wheel in the front of the Holt crawler tractor was virtually the same as the wheel designed by Remington for his early, three-wheeled steam tractor. The rod extending from tractor out toward the wheat aided the driver in knowing just how to position his machine so the combine header would get the maximum cut of grain.

Courtesy, Al Herman

CLETRAC PULLS. The Cleveland Tractor Co. manufactured a track layer they called the Cletrac. Ernest Templin used a Cletrac to pull his combined harvester in 1935. It was originally a horse-drawn machine but was converted to the tractor by adding the stub tongue. Templin and his sons plus another worker supplied the labor force during this harvest near Ritzville, Washington.

Courtesy, A. M. Kendrick

REMOTE CONTROLLED COMBINE. The cat driver didn't need any help to run this combine. He could control the combine motor, raise or lower the header and dump a tankful of wheat into the bed of a truck from the seat of his D4 cat. He could do this with the assistance of hydraulic controls.

Courtesy, Wayne Doty

IN THE BULK. Charley Pierson was one of the early wheatmen to handle his grain in the bulk. Taken in 1921 on his farm near Starbuck, Washington, Pierson is in the driver's seat of his bulk wagon.

Courtesy, Ted Worrall

NEXT STOP: WAREHOUSE. Hauling grain from the Bert Mader farm in the Clear Creek district near Colfax, Washington. The date was around 1910. Mrs. Epler, an early camera enthusiast, took the picture.

Courtesy, Mrs. Glen Epler

MOVING WHEAT TO MARKET. Martin Hansen had five teams of six horses each hauling wheat to the warehouse when the photographer took these pictures. The first shot was taken with the teams and wagon strung out in single file. Then the teams behind moved up, the photographer crossed over to the other side of the road and another photo was snapped.

Courtesy, Sherman County Historical Society

SIXTEEN MULE WHEAT TEAM. Marion T. Weatherford had this picture copyrighted in 1954. Concerning the picture he wrote the following caption; "This picture was taken three miles south of Arlington on the John Day Highway in the summer of 1923. The outfit, consisting of 16 matched mules and seven wagons, was owned by M. E. Weatherford and driven by L. T. (Monk) Coffman. It hauled wheat from the Weatherford Ranch to Arlington, making one round trip of 26 miles each day and hauling 270 sacks of wheat. To aid in making the four right-angle turns in Arlington, two "Kar-Kouplers" were used. Four of the seven wagons had brakes and "Skinner" Coffman always rode the near wheeler, using only the jerk-line and spoken commands to control the near leader. Because of the mental strain on the jerk-line leader, two animals were used, each on alternate days, allowing a day of rest between trips. Note the bells, shielded by angora goat housing, on the leaders. All 16 mules are in step with the ringing of the bells."

Courtesy, Marion T. Weatherford

146

DUST OR MUD. The Condon to Arlington road in eastern Oregon was a busy one when farmers began moving their wheat to the river port. As long as it was dry the road was ankle-deep in dust. Close inspection of the wheels on the first wagon shows how they've sunk into the powdery dust. If rain started falling the dusty path quickly turned into a quagmire of mud. This scene was approximately 1910.

Courtesy, Ted Worrall

FARM WAGON AD. Studebaker was one of the main manufacturers of farm wagons. Many farmers used the Studebaker wagon to haul their wheat to town.

Kirby Brumfield

HEAVY TRAFFIC. Wheat-hauling traffic was heavy on the grade between Clarno and Shaniko, Oregon. The teamsters drove through Shaniko north to the Columbia river where the wheat was transferred to boats for shipment downstream to Portland.

Courtesy, Joe Johnson

PILING SACKS. There wasn't an inch lost when it came to packing wheat sacks into ware-houses. Even so, there was never enough room inside. The surplus was piled outside. The figure "1910" written on one of the sacks could refer to the year but more likely indicates the number of sacks belonging to one of the farmers. The initials designate which sacks belong to each wheat grower.

Courtesy, Sherman County Historical Society

STACKS OF SACKS. It's easy to guess how much labor was eliminated when farmers started bulking their wheat. Each of these sacks was jigged, sewed, picked up from the field, loaded onto a wagon, unloaded at the warehouse, stacked in the warehouse and loaded onto railroad cars by hand labor. Bulking eliminated much of the muscle strain associated with sacked wheat.

Courtesy, Sherman County Historical Society

WAREHOUSE TIE-UP. It didn't do any good to get riled about the long wait to unload at the warehouse. It was better for the digestion to just sit back and relax. One by one the wagons were unloaded and eventually each driver had his turn even though the wait sometimes dragged on for what seemed like hours.

Courtesy, Charles Freeburg

WAITING THEIR TURN. The block behind the right rear wheel was attached to the wagon by a chain. It was a simple but effective way of keeping the wagons from rolling backwards.

Courtesy, Sherman County Historical Society

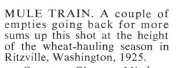

MULE TRAIN. A couple of empties going back for more sums up this shot at the height of the wheat-hauling season in Ritzville, Washington, 1925.

Courtesy, Clarence Mitcham

MULES DELIVER WHEAT
SACKS. There were times when
eight mules could pull three
wagons of wheat sacks. When
the distances were long though,
and the roads rough, rutty and
muddy they were hooked up to
only two wagons.

Courtesy, Sherman County
Historical Society

NO MORE ROOM IN THE WAREHOUSE. Once the farmers began hauling their wheat the warehouses were deluged with more grain sacks than they had capacity to store under roof. If they couldn't stack them inside they stacked them outside. This warehouse at Heppner, Oregon was virtually surrounded by piles of sacked wheat.

Courtesy, Oregon State University Archives

WHEAT EMPIRE. "Sherman County, a Little Wheat Empire" was the title of the lead article in the August 19, 1920 edition of the Oregon Farmer. In it F. L. Ballard, assistant county agent leader for area East of the Cascades, pointed out that, "wheat yields run on an average of from 15 to 20 bushels per acre in Sherman county but yields of 35 to 40 bushels are not uncommon on the better soils. Some seasons the county produces over 2,000,000 bushels of wheat.

The article by Ballard went on to say, "Sherman county is a winter wheat district, the winter varieties returning in most years from 15 to 20 per cent larger yield than fall wheat. Turkey Red is the principal winter wheat and is, in fact, grown almost exclusively from Grass Valley north to the Columbia River. But from Grass Valley south, except around Kent, there is considerable Fortyfold grown," wrote Ballard.

This scene at the Klondyke, Oregon warehouse leaves little doubt about the abundance of wheat grown in the area. The warehouse itself was all but engulfed by the huge, barnlike stacks that had to be built until trains came in and started hauling it away. They were still piling more grain in when this picture was taken as seen by the man with the handcart at the left.

Courtesy, Sherman County Historical Society

TRAMWAY TRANSPORTS WHEAT. From 1879 to 1887 wheat was chuted in pipes from the top of the hills above the Snake river down to the river level for steamers to haul, according to Rex Weiler, former steamboat deckhand. The method was unsuccessful because it burned and scarred the wheat. The next step was to build tramways down the steep slopes. This was the Mayview tramway which ran from 1891 to 1943. It was located about 28 miles below Lewiston, Idaho. The tramcars carried about 40 sacks apiece down a drop of 1500 feet. They moved from three to four thousand wheat sacks down to the river every day. The cars passed half way up the incline.

Courtesy,
Edward Crane Francisco

STEAM ON THE SNAKE. There used to be steamboats on the Snake river as well as the Columbia. One of the chief products they hauled was wheat. According to Rex Weiler of Millwood, Washington who used to work on these boats, there were three "Lewistons." The first was built in 1867. The second was built in Riparia, Washington in 1894. It was 513 tons and 165 feet long. In 1905 it was rebuilt even larger until it weighed in at 548 tons and measured 166 feet in length. It burned in 1922 said Weiler. He remembers the Lewiston could take 1800 sacks of wheat up the river to Lewiston, Idaho but could carry two to three thousand sacks down stream. There were usually about 16 sacks to the ton. Loading the steam-powered river queen was quite an art. If the cargo was placed too far back the sternwheel rode so deep that the wheel was in danger of hitting rocks.

Courtesy, Clarence Mitcham

WHEAT ON THE BANK. These sacks of wheat were piling up along the banks of the Columbia river waiting to be shipped downstream to Portland. This shot was taken by Benjamin Gifford, a commercial photographer from The Dalles, Oregon, and bears a copyright date of 1899. The board stringers extending out from under the sacks were used to keep the sacks off the ground. If set flat on the ground the sacks would likely have picked up moisture and a portion of the grain would have spoiled.

Courtesy, Sherman County Historical Society

GRAIN ELEVATOR. Most of the wheat was stored in warehouses in the early days because it was handled in sacks. By 1918 though when this photo was taken there was a definite move to "bulking" the grain. Once bulking started the shift was on from warehouses to elevators for grain storage. This 100,000 bushel concrete elevator was at Athena, Oregon. It was one of six such elevators owned by the Farmers Union Grain Agency of Pendleton.

Courtesy, Oregon State University Archives

BULK HANDLING. This wheat dumping scene was taken by Curtis at Asotin, Washington, August 10, 1939. The elevator was one belonging to Lewiston Grain Growers Association.

Courtesy, Washington State Historical Society

Chapter VI

Modern Harvest...

JUST AS MANPOWER REQUIREMENTS dwindled with the introduction of the combined harvester, the demand for horses and mules went drastically down as gas tractors were developed. Farming was now entering an era that would lead to total emphasis on mechanization.

As progress took hold of the wheat harvest the curtain closed on the day when it required an army of men to get the job done. The economics of the machine revolution dictated a direct relationship between fewer men and more profit.

Today the harvest crew has dwindled from twenty or so men to a mere three. One man drives the combine and the other two chauffeur trucks back and forth between the field and the grain elevator. There's no need for sack-sewers nor sack-jigs since sacks aren't used. The wheat is handled in "bulk." The combine augurs the grain it has just threshed into an attached holding bin where it stays until emptied into a truck.

Header-punchers, teamsters, straw-bucks, steam engineers, firemen, hoe-down men, roustabouts, and stackers are no longer jobs, they're merely words in a fast-fading vocabulary. In fact, these titles would prompt a puzzled look from many present-day farmers.

Wheatmen no longer harangue over the comparative values of the horse and mule since the last 33 horse and mule teams passed from the scene long ago.

Old-timers will tell you that wheat harvesting today doesn't carry with it the color and romance of those early days. The $20,000 air-conditioned combine of this generation is more efficient and less dusty for the operator perhaps, but to many a poetic heart the wheat harvest is now too dehumanized. One of the people who harbors this feeling and doesn't mind saying so is Giles French, editor of the Sherman County Journal in Moro, Oregon. He editorialized on the subject in a March 28, 1941 column when he wrote a piece titled, "One Hundred Acres a Day." Here's what he said:

"They plow a hundred acres a day now, that is, day and night; great, brightly painted, dust-covered behemoths, like army tanks, go charging over the landscape dragging along strings of plows behind them. A hundred acres a day.

"Time was, and not too far distant, when they plowed with six-horse teams—little wiry horses with bushy manes and tails and the habit of bucking the harness off when it rattled—and five acres a day was considered enough for any man or any team.

"And Saturday night or Sunday the whole crew went to town, the boss and his wife occupying the seat of the hack, and the children and the hired man filling the back. The barbershop was full of animated humanity, the saloon resounded to joke and song, and the store clerks were busy until all hours.

"In the still dark or in the bright spring afternoon, they came back again to the ranch to feed the stock and change plowshares. And there was candy for the littlest ones and a fresh supply of tobacco for the men.

"Now the men who turn over fifty acres a shift are lonesome men. They sit a long ten hours, twelve hours, listening to the rattle of the motor; they grab their lunch at noon or midnight between gear shifts and have communication with neither man or beast in their work.

"Farm women used to cook home-grown food for a table full of dusty teamsters and took the full day; now they serve fine fare to a few mechanics and never miss a bridge game.

"But where is Bud, who used to get his team out first every morning, keep them fat and never have a sore-shouldered horse; or Joe, who always had a song and who had so much fun on Saturday nights; or Pete, who was saving his money to marry the girl over on the next ridge? What do they do to earn the bread they eat?

"One man now raises many times the bushels of wheat he did in those days. There is more efficiency, but no one is happier and there is no life around the town. Farming that was once a way of life, albeit a hard, long-houred way, is now a business, geared to the machine, as surely as any factory; and the raillery of the crews has been stilled by the rattle of valves.

"A hundred acres a day."

STUBBLE MULCH. The moldboard plow is frequently used today in summer-fallow areas. It turns the earth and yet leaves a portion of the stubble on top of the ground. This is useful in fighting wind and water erosion. The exposed straw tends to anchor the soil. When first promoted the practice was called "trashy fallow." Since the word "trashy" had a negative appeal the name was changed to "stubble mulch."

Courtesy, Washington Association of Wheat Growers

ROD WEEDER. The rod weeder was doing an effective job of keeping the weeds down on this summer follow operation near Ritzville, Washington in 1965.

Courtesy, Washington Association of Wheat Growers

NIGHT LIFE. There are times when the wheat harvest goes on even after the sun sets.

Courtesy, Washington Association of Wheat Growers

BIG INVESTMENT. It takes money and lots of it to get started in the wheat business. The average wheat farmer has an investment of at least $200,000 in his operation. Land and machinery take tremendous outlays of capital. This partially explains why most wheat men are descendants of the pioneers who homesteaded the land originally. Since the return on investment seldom measures up to what a businessman in town would expect it is not economically feasable for a person to go out and buy a ranch and start farming—even if there were land available. Ed Martin of Walla Walla, Washington stands beside a small portion of the equipment he needs to crop wheat.

Courtesy, Washington Association of Wheat Growers

BOOKKEEPING. The successful farmer bases his operation on an up-to-date set of books. Often the wife is a big asset in this area.

Courtesy, Washington Association of Wheat Growers

MODERN PLOW PLOWER. Plowing today is a lonely, one-man job.

Courtesy, Washington Association of Wheat Growers

Chapter VII

Wheat Grower Associations...

NORTHWEST WHEATGROWERS have always been a hardy lot. They got that way from fighting bad weather, poor crops and short labor. Each farmer faced up to these problems with grim-faced determination. His set jaw signified an almost undestructible will to survive. There were irritations and discouragements but the grain men gamely reminded themselves that every job had its hazards.

When times were particularly tough and wheatgrowers found it an uphill pull to keep going, they made it on guts alone. As one farmer phrased it, "I had reached the end of my rope so I tied a knot and hung on a little longer. It worked out."

Wheat farmers found themselves in a particularly bad way during the depression that followed World War I. Their wheat was selling for less than it cost to grow.

There's little doubt these wheatmen were hard-headed but no one can ever accuse them of being thickheaded. Independent and strong-willed as they were, they realized there was strength in numbers.

Under the guidance of the Oregon Agricultural College Extension Service, growers from the major wheat-growing counties of eastern Oregon gathered for a historic meeting at Moro, Oregon, February 11-12, 1926. It was called the Wheat Growers Economic Conference.

The main purpose of the session was to take a penetrating look at the wheat business, what it was doing and where it was going. More than one hundred leading wheatgrowers served on five committees along with bankers, transportation men, millers, elevator and warehouse men, experiment station and extension service workers and representatives of the United States Department of Agriculture.

Each committee was assigned a definite topic. The committee titles were farm management, finance and credit, tillage and production, wheat handling, and world supply and demand.

So often meetings turn out to be one of three things: a social function; a gripe session; or a place where people pool their ignorance. The Wheat Growers Economic Conference was none of these. The participants were looking for facts, and facts they got. It could well be that these farmers gathered around tables laden with more facts and figures relevant to their industry than had any group of producers in previous history.

Key fact suppliers were: L. R. Breithaupt, secretary of the World Supply and Demand committee and member of the extension service at Corvallis who had spent a month in Washington, D. C. tracing down all the helpful information he could find; Byron Hunter of the Bureau of Agricultural

Economics of the U.S.D.A. who assembled volumes of information and had many charts made; Dr. W. J. Spillman, consulting economist from the Bureau of Agricultural Economics, U.S.D.A., Washington, D. C., and formerly of Washington State University who presented much valuable data.

Also on hand were numerous staff members of Oregon Agricultural College including H. D. Scudder, head of the farm crops department; R. S. Besse, farm management demonstrator; G. R. Hyslop, later to be head of the farm crops department; E. R. Jackman, crops specialist; county agents and many others. F. B. Ingels of Dufur was elected chairman of the conference.

IMPORTANT REPORT. The Wheat Growers Economic Conference held in 1926 at Moro, Oregon resulted in the formation of The Oregon Wheat Growers League. This report was packed with facts and recommendations to help wheat growers meet the challenge of depressed prices.

Courtesy, Sherman County Agricultural Extension Service

As Wallace Kadderly of the O.A.C. extension service reported in a bulletin summarizing the meeting, "There was no speech-making at the conference. Those who attended joined the committee in which they were most interested. Two days were spent in considering information gathered and in arriving at recommendations. Everyone had an opportunity to express his views. Each group then put its recommendations into a report that was considered by the joint session of all groups on the last day and adopted."

The farm management committee dug into the costs of producing wheat. They concluded that, "The actual cash out-of-pocket cost of wheat production is about half the total cost. The average total cost of producing wheat for the years 1922, 1923, and 1924 on land having an average value of $50.00 per acre is $25.00 per acre, of which approximately 50 per cent or $12.00 per acre is the cost in cash and $13.00 per acre is the non-cash cost.

"Non-cash includes items of wages for the operator's own time, value of food for horses, value of seed used, depreciation of machinery, buildings, and stock, and interest at six per cent on all of the investment not covered by mortgage. When the price received for his wheat equals the total cost, the grower gets cash payment for all items of cost, both cash and non-cash items, and wheat growing is a good business." Farm economists today might not be in complete agreement with that statement.

One of the more interesting subjects under discussion concerned itself with horses and tractors and which was the most practical to use. It may seem surprising that horses were even considered at a forward-looking economic conference as late as 1926. What may be even more jarring is the fact that horses received strong support as the most economical. One of the reasons they were considered more practical was their surefootedness on steep eastern Oregon sidehills. Tractors, big and awkward, were still in a development state.

It was more than that though. Again, the decision was based on facts, not emotion. The committee backed up its recommendation with a cost-conscious chart showing why the work horse still had a place. It included the following information.

TRACTOR-HORSE COST COMPARISONS
BETWEEN TWO GROUPS OF FARMS
OF THE SAME SIZE

One group operated by tractors and horses. The other group operated by horses alone. Data from wheat cost survey—Sherman County.

THE OLD AND THE NEW. A Case 110 steam engine stood next to a current Model 1200 King tractor during a Steam-Up at Rockford, Illinois in 1966. In contrast to these two giants is the little Case compact tractor used mainly for garden work. Steam-Ups and Threshing Bees have become extremely popular around the country. It affords old-timers an opportunity to hash over accomplishments of harvests gone by. It also gives them a chance to actually work the machinery that did the job a half century ago.

Courtesy, J. I. Case Co.

Average of three years	*25 tractor farms*	*40 horse farms*
Size of farm—acres	1183	1230
Acres of crop and summer fallow	924	905
Number of horses per farm	14.6	20.6
Value of horses per farm	$1704	$2247
Value of horses per head	$ 117	$ 109
Cost of keeping horses per farm per yr.	$1424	$1741
Cost of keeping horses per head per yr.	$ 97	$ 85
Hours worked per head per year	649	830
Cost per hour of horse labor	$ .15	$ .10
Cost of all power per acre	$ 3.36	$ 1.92
Cost of power and man labor per acre .	$ 7.00	$ 4.91
Total net cost per acre of wheat	$ 32.87	$ 26.77
Total net cost per bushel of wheat	$ 1.34	$ 1.06
Percentage return on total farm invest.	4.4%	5.6%

Based on this chart and others the conference committee made the following recommendations on the horse versus tractor issue:

1. Under average conditions the wheat farms having less than 1000 acres of cultivated land per farm produce wheat at less cost per acre and per bushel with horses alone than with tractors and horses and hence find horse-power operation more profitable than tractor-power operation.

2. Under average conditions wheat farms having more than 1000 acres in cultivation produce wheat as cheaply and in many cases at lower cost per acre and per bushel with tractor and horse operation combined, than with horses alone.

3. Even on large farms horse operation is efficient and need not necessarily be converted to tractor operation.

When it came to the decision between combines or a header-thresher operation the conclusions were these:

1. While each have their advantages, under average conditions harvesting wheat with the combined

HOTEL ON STEPTOE BUTTE. James (Cashup) Davis had a dream of a mansion on a hilltop. The hilltop was Steptoe Butte. He decided to build a hotel on top of the mountain and have it be the greatest showplace in the area. It was an immense dream but Cashup had a way of making dreams come true. Unfortunately, the hotel was a failure. Built in 1888 it had a novelty appeal that drew sparse crowds for several years. Soon, though, the novelty wore off and Cashup Davis had the place all to himself. He stayed there and died a broken-hearted man in 1896. He was 81 years old. One of his boys tried to keep it going for a short time. It burned down in 1911.

This picture was taken in happier days. Old-timers estimate the year was 1889. The inscription on the picture says, "Steptoe Butte 3800 Feet High. THE BEST PLACE IN WASHINGTON TO GET A GOOD VIEW OF THE COUNTRY." A second piece of writing says, "WE ARE WAITING FOR THE CARS." Randall Johnson, leading historian on the subject, says "cars" referred to the cars on an excursion train which Northern Pacific railroad promised to run in the valley below. They made good on their promise but there just weren't enough people taking the train to continue the run.

Courtesy, Bill Walters

harvester costs less per acre and per bushel than harvesting with header and stationary thresher. Data from the Wheat Cost Survey show that it cost 95 cents more per acre or 6.2 cents more per bushel, for the harvesting and marketing of wheat harvested with headers and stationary threshers than for wheat harvested with combines.

2. Motor-driven combines harvest wheat at less cost than ground-power combines. Data from the Wheat Cost study show that while the cost of ground-power combines is much less per day of use than for motor-driven combines, more horse-power is required for them, much less wheat is harvested per day, and inferior work is done. These machines are now regarded by most farmers as obsolete.

3. Horse-drawn combines harvest wheat at less cost than tractor-drawn combines.

The tillage and production committee compiled a list of the most successful cutural practices, best varieties, rates and dates of seeding based on the experience of leading eastern Oregon wheat growers and the Moro Branch Experiment Station.

The 1926 recommendations were:

1. Stubble should never be burned in the fall. The stubble aids in holding snow and in moisture absorption. The turning under of any form of crop residues will, in the long run, likely prove profitable because of the inherent lack of humus and nitrogen in Columbia Basin dry-farm soils.

2. Disking stubble ground in the fall has a tendency to hinder moisture absorption and to reduce wheat yields.

3. For early spring plowing, disking does not pay. Whenever possible jointers should be used to turn under stubble more completely.

4. Late plowing without previous disking reduces wheat yields.

5. Thorough spring disking on certain types of light soil may occasionally replace plowing, but for the majority of soils no tillage implement yet invented will take the place of the plow.

6. Plowing from five to eight inches deep with variations in depth each time of plowing is recommended.

7. Packing after plowing does not materially affect the yield of wheat grown after fallow.

8. Harrowing should be done within a week or ten days after plowing.

9. Tillage tools best adapted for cultivating summer fallow are: spike-tooth and spring-tooth harrows, and blade or rod weeders.

10. All weeds should be kept off the summer fallow. Weedy fallow means lower wheat yields and poorer quality of wheat.

11. If the seed-bed is not smooth, the harrow should precede the drill, even when sowing is done in dry ground, except on blow soils.

12. The best rate of sowing winter wheat is between 3 and 5 pecks per acre. The rate should be varied according to moisture conditions, time of sowing and condition of seed-bed and size of kernel. In Jefferson county and certain section of Union county, thinner seeding is advised.

13. For most sections of eastern Oregon, higher winter wheat yields can be obtained from comparatively early sowing; i.e., from September 15 to October 15. Later seeding may be advisable for shallow soils.

14. Winter wheat should be sown shallow, ordinarily one to two inches.

15. Winter wheat varieties recommended for general culture in eastern Oregon are; Turkey, Hybrid 128, and Fortyfold. Caution should be exercised

WHEAT DECOR. The influence of wheat in the area was certainly brought out in the interior design of the hotel. Wheat was everywhere—hanging from the ceiling and sitting about in pots, pans and cans. The man with book on his knee was an itinerant preacher named Sproat. The book may likely have been a Bible. Cashup Davis knew all about wheat. He had grown it for years, being one of the earliest pioneers in Whitman County, Washington. In a sense, the wheat-inspired decor was Cashup's way of paying tribute to the plant that had been so good to him.

Courtesy, Bill Walters

MPER CROP
1937
ITZVILLE, WN.

13,000 SACKS

A MOUNTAIN OF WHEAT. Ritzville, Washington has always been a key wheat center. In 1937 its bins were bulging once again. The 13,000 sack pile was built in the perfect shape of a warehouse shed.

Courtesy, A. M. Kendrick

in the planting of Federation in the fall because of its lack of winter hardiness. Recommended spring wheat varieties are: Federation, Hard Federation, and Baart.

16. Spring wheat should always be sown early. The most profitable rate for Federation is 3 to 5 pecks; for Hard Federation and Baart 4 to 6 pecks.

17. Use the same variety as the rest of the field for hay strips if possible. Sow a white wheat if field is a white wheat; and a red wheat is field is a red wheat.

18. The use of copper carbonate is strongly recommended for treating wheat for smut.

19. Always sow good, clear seed, free from weed seeds.

20. Grain certification work has been of value. It should be continued.

21. Farmers are strongly advised to use every possible precaution to keep summer fallow free from weeds.

22. For the control of morning glory, the application of salt is the best remedy yet found for small patches. For larger areas, clean cultivation is the most practical method of eradicating this noxious weed.

23. The general or extensive culture of other crops on the typical wheat lands of eastern Oregon is not recommended. Under certain conditions and in certain localities, field-peas and corn, to a limited extent, may be profitable for farmers who keep livestock. In Union county, and in some sections of Wasco and Umatilla counties, alfalfa, sweet clover, corn, and peas are profitable crops. On limited areas in Union and Umatilla coun-

ties, potatoes and beans are also profitable. We strongly recommend the keeping of enough livestock and poultry on every wheat farm to utilize advantageously all by-products, and to make use of land not suited to profitable wheat production.

This has been just a brief look at some of the conclusions of the farm management and the tillage and production committees. The participants waded into the other three committee assignments with the same thoroughness. Except for a spontaneous Lincoln Day speech from Walter Pierce, leading farmer and future Oregon governor, the economic conference spent every spare minute analyzing their industry.

The Wheat Growers Economic Conference, called into session in a troubled period, proved a springboard to the formation of one of the most effective commodity organizations ever developed . . . the Oregon Wheat Growers League.

The 1926 meeting had been such a help to the farmers they decided it should be an annual affair. Elected president was Frank Ingels of Dufur. A meeting was slated for the next year, 1927. However, Ingels became ill and passed away before the meeting took place. As a result, the second annual meeting didn't actually occur until 1928. It was at this get-

NEW MARKETS. With an eye to increasing their overseas market, wheat farmers of Oregon, Washington and Idaho organized the Western Wheat Associates. Wheat was an unfamiliar grain to many countries such as Japan, India, East Pakistan and Taiwan but the WA has been extremely successful in introducing the product to consumers and processors in these lands. Food fairs, workshops and demonstrations have been used to accomplish the task. Demonstration kitchens-on-wheels have taken the story of better food through wheat right to the people. In the city or out in the country there are always eager onlookers who crowd around the buses or booths to get free samples of the new and tasty foods. The home economists also dispatch ample helpings of information on nutrition and sanitation as they talk about food preparation.

The success of this program has opened up a large, previously untapped market for the wheat of the Northwest. The accompanying photos supplied by the Washington Association of Wheat Growers show some of the activities in these countries that resulted from the action of farsighted wheat farmers in Oregon, Washington and Idaho.

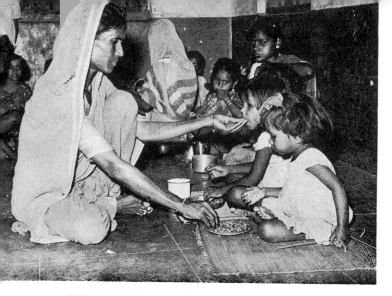

DINNERTIME. A mother of eleven at a YWCA sponsored feeding center in Bombay, India is feeding bulgur wheat to her youngest children.

Courtesy, Washington Association of Wheat Growers

together that the official name of Oregon Wheat Growers League was adopted. From the beginning down to the present, their basic approach has been "What can WE do to solve our own problems?"

Frank Ballard, former director of the agricultural extension service for Oregon, has a quick response when asked about the Wheat League. He says, "It became the most active agricultural organization in the state. The influence of the League has been tremendous. The group is still the most studious and best informed on their business of any in the state. The annual meetings have been times when the farmers dug deep into the undercurrents of the wheat economy."

On a nation-wide scale, the National Association of Wheat Growers had its birth in Kansas City in 1951. Much of the inspiration and leadership for this organization came from the already proven Oregon Wheat Growers League. A parallel organization was called the Washington Association of Wheat Growers.

Oregon was also the first state to form a wheat commission. A booklet titled "They Broke the Trail" tells of a meeting of Northwest farm organizations gathered in Pendleton in June, 1945. The topic was federal programs. According to the booklet, "Mr. H. R. (Ike) Weatherford, at tht end of the long discussion asked, 'Why don't we put a tax on wheat and do some of these jobs ourselves?'

"This self-help idea appealed to these Oregon wheat growers and in 1946, the Oregon Wheat Growers League recommended a wheat commission be formed. Jens Terjeson of Pendleton, and Marion Weatherford, Arlington, were given the job. With characteristic thoroughness, they spent more than a

month at Salem during the 1947 session of the Legislature buttonholing legislators and explaining just what was wanted. The Oregon Wheat Commission bill was passed with only one dissenting vote."

Financing for the commission came from the farmers themselves. They paid one-half cent for every bushel of wheat sold. This gave the group a fairly sizable budget to begin their new program.

Until this time no state in the country had a wheat commission. Since the Oregon Wheat Commission was the first of its kind there were no set patterns to follow. It was like venturing into unexplored territory.

Once again from the booklet we read, "After the Commission was organized, breaking new trails became commonplace. The admonition in the law to 'Conduct a campaign of research, education and publicity to promote wider markets for Oregon wheat' was taken literally. The key words are 'wider markets,' and 90 per cent of the Commission's activity has been geared to this one goal."

Just one example of how the Commission has been effective in expanding the market for wheat was its work with a product new to the average American housewife. This new product has also been of tremendous importance in wheat trade with Asiatic countries.

This food called bulgur, although relatively new to Americans, has been known to central Europeans for hundreds of year. Bulgur is the term applied to wheat that has been cooked, dried and processed to remove a portion of the kernel.

Mr. John Korenian of Salem, Oregon, told the commission about the tasty wheat product. Later, one of the commission members discovered Omar Khayyam's restaurant in San Francisco featured the dish. Thinking began to jell that surely here was a wheat product that might be popular with American cooks and could be exported.

With the Commission serving as a catalyst, arrangements were made between the Commodity Credit Corporation and the Fisher Flouring Mill of Seattle, to manufacture bulgur wheat on a large scale. Much research had to be done. Both the Fisher Flour Company and the U.S.D.A. Western Regional

HOME ECONOMICS DEMONSTRATION. Eager onlookers snap up new cooking ideas which involve wheat products.

Courtesy, Washington Association of Wheat Growers

Laboratory in Albany, California went to work. The result is a product now marketed in the Pacific Northwest under the trade name of Ala.

Thousands of tons of Ala have been exported to Korea where it is accepted as a standard food item. It is also being used extensively in India, Japan and other Asiatic countries. It is popular with Northwest homemakers and may eventually sell nationally.

A wheat commission similar to Oregon's was established in Washington in 1958. The Washington wheat growers are assessed one-quarter cent per bushel. Idaho followed suit in the spring of 1959 with the formation of the Idaho Wheat Commission. These three commissions have served their farmers in many ways ranging from seeking better freight rates and financing research to developing new markets and encouraging soil conservation.

Washington, Oregon and Idaho pooled their monies and desires for increased markets on May 2, 1959 by forming the Western Wheat Associates—U.S.A., Inc. Its main direction is aimed at developing and maintaining foreign markets for Northwestern wheat. Something like 85 per cent of Oregon and Washington wheat is presently being marketed in such Far East and South Asian nations as Japan, India, Pakistan, The Philippines, Burma and many others.

There's no argument that wheat growers are and have been strong-minded, gritty "individualists." Never willy-nilly in tackling problems, Northwest wheatmen have turned sagebrush and sand into thriving wheatlands. They've fought jackrabbits, disease and low prices. In those early days they fashioned a livelihood out of wheat on muscle and guts alone. Each of these early-day farmers could have been the center of an epic novel depicting man's insurmountable will to make something of nothing.

Fortunately, they also saw the wisdom in uniting.

FOOD BOOTH. Two Japanese farmers enjoy pancake sandwiches in front of Wheat Associates' pancake booth at the Kiyosato County Fair. WA staff member, Toshio Hannya (center), directed the booth, which was aimed at popularizing pancakes and other U.S. wheat products in rural Japan.

Courtesy, Washington Association of Wheat Growers

NUTRITION BUS. One of the best ways of showing people how to use wheat is to put a demonstration kitchen on wheels. This bus used in Okinawa has large doors in the back which drop down. As the people gather around, home economists prepare different foods and relate important nutritional information. The overhead mirror aids the onlookers in seeing exactly how the job is done.

Courtesy, Washington Association of Wheat Growers

The organizations these individualists formed have been of immeasurable value in prospering the product most dear to their hearts . . . wheat.

RESEARCH TO THE RESCUE

Wheat has gained its ranking as number one farm crop of the Northwest in spite of weeds, pests and disease. These sly villains have combined forces in an all-out attack on the popular grain.

By all logic, wheat should have long ago been reduced to a plant of minor importance by its persistent enemies.

On the farmers' side though has been a small army of dedicated agronomists, entomologists and pathologists. Some of these specialists are employed by private industry but most have worked under the banners of land grant universities and the United States Department of Agriculture. Washington State University, Oregon State University and the University of Idaho long ago saw the wisdom of working cooperatively to improve wheat quality and production.

WHEAT PRODUCTS. A cameraman in Taiwan caught two young helpers dishing out bulgur porridge and fried steamed bread.

In some ways the most unenviable job in the world is that of a plant scientist. He does not know what it is to taste absolute and complete success. A wheat variety has never been developed that is totally resistant to all pests and diseases—and probably never will be.

A top yielding wheat plant today will by comparison be a mediocre producer tomorrow. The search for better plants will never cease.

It takes years to develop a new variety. After this comes more years of extensive testing to make sure the newcomer has all-around talent. Of the thousands of varieties developed only a few survive the severe screening tests.

To make it big a new wheat must have triple-threat ability. It must outyield its predecessors, be more vigorous in resisting onslaughts of disease and pests, and stand at the head of the class in milling quality.

The plants that display potential for greatness are given final tests at regional labs such as the Northwest Wheat Quality Laboratory and the Regional Smut Laboratory, both at W. S. U. in Pullman, Washington and at university-operated experiment stations scattered throughout Oregon, Washington and Idaho.

The few varieties that make the grade must experience birth pains extending over a 10 to 15 year period.

Wheatmen are the first to get the word on the latest varieties, weed and insect control chemicals, seed treatments, cultural practices, and economic factors thanks to the agricultural extension service. State specialists and county agents feed up-to-the-minute facts and information to farmers as soon as researchers make a new advance in the technology of wheat.

Northwest grain growers have never been slow or shy about adopting new ideas. They snap at them eagerly. That's why the industry has advanced at a gallop.

WHEAT SHIP. Ships such as this one transport millions of bushes of wheat throughout the world each year.
Courtesy, Washington Association of Wheat Growers

INSPECTING A FRESHLY BAKED LOAF of Japanese bread during a visit to a Yokahoma bakery is Dr. Mark Barmore, Head of the Western Wheat Quality Laboratory at Washington State University, Pullman. Dr. Barmore recently completed a two-week stay in Japan to study the usage of U.S. wheat in this important cash market. His trip was made at the request of Western Wheat Associates, the overseas marketing organization of Washington, Oregon and Idaho wheat producers. Looking on are Mr. Michio Uchida, assistant director of the Japan Institute of Baking, and an unidentified Yokohama baker.

SCHOOL LUNCH MENU. The menu of the school lunch program at Takanawadai Primary School, Tokyo, includes one loaf of bread, a bowl of vegetable soup, a bowl of reconstituted dried skim milk, and one pat of butter.

Courtesy, Washington Association of Wheat Growers

PRODUCTIVE SLOPES. Smut has been only a minor problem in recent years. Scientists teaming together have dealt the troublesome fungus a crippling blow. There's no guarantee it won't bounce back through and that's why researchers are cautious to say the problem is completely licked. This sceen was taken on the Harry Timpey farm near Dayton, Washington in 1963.

Courtesy, Washington Association of Wheat Growers

Chapter VIII

Smut...

ONE OF THE BIGGEST HEADACHES to wheatmen of the Northwest down through the years has been the troublesome fungus known as "smut." It has always been an uninvited guest that wheat farmers would have gladly done without. It has been every bit as deadly to America's number one crop as is cancer to the human body.

The U. S. Department of Agriculture and Land Grant University scientists have faced a frustrating challenge in fighting the yield-robber. The years of laborious field and lab work needed to develop new strains of wheat are usually nullified fairly soon after a plant's introduction.

Unfortunately, these smut-resistant varieties have never been quite resistant enough. As each new wheat variety went into general use, it did what it was designed to do—resist smut attacks. For the first several years after its introduction, smut would seemingly be down for the count.

Then, suddenly a race of smut would pop up which virtually thrived on the so-called resistant variety. It was then back to the drawing board to breed yet another smut-resistant quality into a new wheat strain.

SMUT HISTORY

It's difficult to say when smut first reared its ugly black head in the Northwest but it was likely soon after farmers started cultivating wheat. Dr. C. S. Holton, pathologist in charge of the regional smut research laboratory at W. S. U., reached into the past and came up with some interesting history on the subject. He made the following report.

"Precisely when wheat smut became a production problem in the Pacific Northwest is unknown. Probably it was prior to 1890, at least in the Palouse area of southeastern Washington, because reference is made to wheat smut in a Washington State Experiment station bulletin dated 1892. That year, in a Farmers' Institute program conducted by the experiment station staff at Garfield, Washington, a Mr. Williams asked how to prevent smut in his wheat.

"In the ensuing discussion, another farmer, a Mr. Clark, spoke of smut in his wheat crop, saying, 'A corner of my field on a northwest hillside was badly infected with smut and I charged it to a thundershower which passed over that portion in full fury."

"At a similar institute the following year, 1893, the problem of controlling wheat smut was again brought into the discussion. Curiously enough, however, no further mention of wheat smut was made in the experiment station reports until after the turn of the century.

"In 1902 an experiment station worker named Beattie published a detailed account of wheat

SMUTTY WHEAT. The black cloud pouring out behind the combine was a familiar sight in 1955 and 1956. Those were the worst years for smut infestation in Northwest wheat fields. The black spores cost farmers millions of dollars.

Courtesy, Dr. C. S. Holton, U.S. Dept. of Agriculture, Pullman, Washington

smut, calling attention to its economic importance for the first time by official publication. Even so, it was not until the second decade of the new century that the smut problem was recognized as the foremost wheat production hazard in the Palouse and adjacent areas.

"Although yield reduction was the primary economic factor inherent in the smut problem, it was the smut explosions and resulting fires in threshing machinery that dramatically emphasized the seriousness of the situation.

"At about the same time, other aspects of the problem came to light, including the soil infestation phase of its life cycle, which prevented effective con-

trol by seed treatment, thereby focusing attention on the need for developing smut-resistant varieties," concluded Dr. Holton.

SMUT HITS ALL-TIME HIGH

The all-time big year for smut was in 1955. That was the year that almost every combine moving through the wheat fields of the Northwest was followed by an ominous black cloud of smut spores.

Based on a release from the Washington State University agriculture extension information office, the Spokane Daily Chronicle of January 11, 1956, printed a story which it headlined, "Smut Takes Record Toll of Wheat in Northwest."

The article made it quickly clear how devastating the smut had been to the wheat crop when it lead off with this statement: "Last year was the blackest smut year on record for Pacific Northwest wheat."

The story went on to say, "In a year when farm income took a 10 per cent clump, the black fungus which replaces the wheat kernel cost Northwest wheat growers an estimated $5,000,000.

"The dollar deficit was totaled in terms of production losses, grain dealers' deductions and washing charges on a total of 29,000,000 bushels of smutty wheat. The total is up 5,000,000 bushels from 1954.

"Elmar, the white club wheat released just five years ago by plant breeders and hailed as resistant to half the known races of smut, registered as high as 83.3 per cent smutty in one section of Idaho and 53.8 per cent smutty over the entire Northwest.

"Of all wheat grown in the Northwest last year (about 88,000,000 bushels) 34 per cent graded smutty, 11.7 per cent more than in 1954. The Northwest of the survey includes Washington, Oregon and northern Idaho."

The picture wasn't any better in 1956. The Washington Farmer magazine reported, "Smut continues to be Public Enemy Number One to Washington—and Pacific Northwest—wheat growers, claiming something like a $5,000,000 tribute again last fall from this area's grain farmers. For 1956, 26.7 per cent of the wheat in Washington, Oregon, and Idaho graded smutty.

"Although fewer fields actually smutted last summer, monetary loss equalled the 1955 toll because infestations were more severe where they did occur.

"These facts and figures on the '56 smut situation were reported by Dr. C. S. Holton, Pullman. Holton's report was based on a questionnaire distributed through the Pacific Northwest Grain Dealers' Association.

"The report showed that smut conditions were worse in Washington and Idaho than in Oregon. In Holton's survey, the Walla Walla area led all the rest in smut incidence with 42.5 per cent smutty wheat. The Palouse was a close second with 38.6 per cent, down 24.4 percentage points from last year's 61 per cent. The Lewiston area was third with 35.3 per cent smut, followed by Pendleton with 34.2 per cent. Spokane had 14.7 per cent smut; The Dalles, 6.5 per cent; and the Lind area, 3.0 per cent.

"By states, Idaho, second in total wheat production with almost 39,000,000 bushels, had the highest percentage of smut—34.9 per cent in Holton's sample. Washington, with the highest wheat production of the three states—almost 60,000,000 bushels—was second with 26.1 per cent. Oregon, with almost 26,-000,000 bushels of wheat, was third in both production and smut percentage—23.4.

"Elmar again led the list in both productive volume and smut percentage. Over half, or 54.1 per cent, of the variety graded smutty. This represents 42 per cent of the survey sample and about 80 per cent of the total volume of smutty wheat.

"Dr. Holton points out that to whatever extent Elmar is replaced by the highly resistant Omar in 1957, the smut problem will be reduced accordingly. All samples of Omar checked this year were smut-free."

In spite of Holton's statement about the Omar variety putting up a completely successful fight against smut in 1956, it wasn't but a few years until Omar too fell prey to the persistent black spores.

To show how short-lived the resistance is in a resistant variety, it is enlightening to read a press release which came out in 1959, The release quoted Dr. J. M. Raeder, University of Idaho plant pathologist, as saying, "New races of smut constantly spring up to attack the resistant wheats we develop. Omar is a good example of a wheat that for a time was entirely resistant to all races of smut. Then we planted so much of it that somewhere a smut race developed to attack it, too," concluded Raeder.

In spite of smut's ability to bounce back, it has not, "knock on wood," been a significant problem during most of the '60's. There's no guarantee it won't again rear its ugly spore-filled head, but for now the threat is seemingly under control. As Anna Jim Erickson wrote in a February 1, 1962 press release from the W. S. U. Agricultural Extension Service information office, "Wheat smut in the Pacific Northwest is down for the count, but not out. And it could bounce back anytime wheatmen drop their guard."

The one-two punch combination which sent smut reeling was chemical seed treatments and the continual development of resistant varieties.

Most of the regionally-adapted new wheat varieties in recent years have come about under the leadership of Agricultural Research Service agronomist Dr. O. A. Vogel stationed at Pullman, Washington. Omar, Brevor, Burt, and the presently popular Gaines were developed by Vogel.

The decline of common smut beginning in 1956 also coincided with the development of improved chemical seed treatments by the USDA Regional Smut Research Laboratory in Pullman. This program headed by Dr. C. S. Holton conducts annual surveys for new smut races, maintains cultures of new races, and studies the life cycle of the fungus. A.R.S. plant pathologists Holton and Dr. L. H. Purdy, Jr., found compounds containing hexachlorobenzene (HCB)

SMUT EXPLOSION. Smut was a damaging and dangerous threat to a farmer's wheat crop for more reasons than one. A smutty crop resulted in a lower grade and lower price paid to the farmer. Of even more concern than price though was the explosive quality of smut. The outfit in this picture burned out because of a smut explosion. It happened in 1915. This picture was printed on a postcard. Someone had written on the back of the card that there was probably 20 to 25 per cent smut in the wheat.

Courtesy, Washington State University Archives

killed soilborne as well as seedborne spores of the smut fungus when applied to the seeds before planting.

The Holton reports on smut prevalence in the late '50s and through the '60s have shown amazingly low amounts of the black fungus compared to the heavy incidence in the mid '50s.

The difference is an eye-popping plunge from the

WHEAT RESEARCH. This researcher is counting smut in a 1923 wheat nursery. He held a tally register in each hand, one to count healthy wheat heads and one to count smutty wheat heads.

Courtesy, Oregon State University Archives

FIELD DAY. Experiment stations scattered throughout the Northwest were outdoor labs in which plant scientists carried on continual research to develop higher producing and more disease resistant varieties of wheat. Each year farmers were invited in and given a report on progress. The best cultural practices were discussed, again based on research. These field days brought out large numbers of farmers eager to learn the latest.

Courtesy, Oregon State University Archives

record toll of 34 per cent smut in 1955 to less than one per cent for the nine years up through 1967. Monetary losses on a regional basis have been negligible in these recent years. That compares most favorably with the estimated $5,000,000 loss to wheat farmers in both 1955 and 1956.

Even though smut is not presently heading the list of troublemakers for Northwest wheat, Dr. Holton still balks at claiming complete victory over the plant disease. He emphasizes that although smut amounts are low, "As long as any fungus spores remain in the area, the explosive outbreaks and multimillion-dollar losses of the mid-fifties could recur."

In spite of this warning, with men like Drs. Holton and Vogel, along with W. S. U. plant scientists Drs.

Calvin Konzak and Robert Nilan who use space-age techniques of radiation to develop new strains, combined with O. S. U.'s Dr. Charles Rohde and Dr. Warren Kronstad in addition to Agricultural Research Service's Dr. Robert Metzger and University of Idaho's Dr. Warren Pope plus many more, it would seem that smut may have hopefully met its Waterloo.

MODEL T POWER. Research in developing better strains of wheat has gone on for years. The major work has gone on for years. The major work has been done by college and U. S. Department of Agriculture scientists. This miniature threshing machine at the Waterville branch experiment station was run by a Model T in 1920. A rear wheel of the Ford ran on rollers connected to a belt which powered the separator.

Courtesy, Washington State University Archives

INSPECTING WHEAT. Professor G. R. Hyslop inspected Vic Roumagoux's field of Blue-stem wheat for certification at Pilot Rock, Oregon the summer of 1921. The crop was grown from certified seed shipped in by Umatilla County Agent Fred Bennion.

Courtesy, Oregon State University Archives

HIGHEST YIELD. Fred De Wilde of Oak Harbor, Washington looks proudly over his prize-winning field of Red Russian wheat in 1919. The yield averaged out at 84.5 bushels per acre. De Wilde received $1,000 in prize money from the Farm Journal magazine for the highest yielding wheat in the United States that year.

Courtesy, Washington State University Archives

WHEAT SILAGE. It wasn't often that wheat in Washington was grown for anything but grain. This picture shows one of the exceptions. These farmers are filling a pit silo with spring wheat. The wheat-into-silage operation occurred in Grant county, Washington in 1915.
Courtesy, Washington State University Archives

WASCO COUNTY SCORES. Wasco County walked off with the first prize and sweepstakes award in county exhibit competition at the 1923 Oregon State Fair. The purpose of such a booth was to display the agricultural products of a county. The large number of wheat sheaves made it plain that wheat played an important part in Wasco county's farming picture. Booths such as this have almost become a traditional part of American folk art.

Courtesy, F. L. Ballard, O.S.U.

UP IN SMOKE. One of the most horrifying sights to a wheat farmer is a grain field on fire. The tinder-dry crop is particularly susceptible to fire along a main road. Careless motorists frequently flick cigarettes out their car windows into wheat fields. This fire ravaged 30 acres on the Walter Jantz farm near Ritzville, Washington. The remainder of his irrigated (wells) Gaines wheat crop averaged 72 bushels per acre. *Courtesy, A. M. Kendrick*

SPRAYING MORNING GLORY. Power sprayers date back at least to 1929. That's when a W.S.U. researcher tested this rig. It was being used to spray sodium chlorate on patches of morning glory. The outfit was mule driven and powered by a stationary gas engine. It carried a 200 gallon tank.

Courtesy, Agricultural Extension Service, Washington State University

COMMUNITY GOODWILL INSPIRES HARVEST SPECTACULAR. No one is quicker to respond to a call for help than a farmer. In fact, Everett Herring of Ritzville, Washington didn't even issue a call for help when neighboring farmers came roaring to the rescue. They knew Ev was sick and couldn't get out to harvest his 440 acre wheat field. Rather than let him fret and worry about it the folks of the community put their combines together and took action. At 7:00 a.m., Sunday morning, July 31, 1966, 13 combines pulled into the vast field and started eating away at the golden bounty in a ravenous manner. The crew took two hours off for a delicious farm-cooked lunch prepared by the wives. Back in the field the men had the last wheat kernel harvested by 4:00 p.m. which must have set some kind of a record. It's doubtful that such a bevy of combines ever tackled a wheat field before or since.

Courtesy, A. M. Kendrick

LARGE BARGE. Much of the wheat from throughout the Columbia Basin is channeled into Pasco, Washington for barging down the Columbia to Portland, Oregon and then shipped to points around the world.

Courtesy, Washington Association of Wheat Growers

THE BIG LIFT. How things have changed. The Cargill wheat shipping facilities in Pasco, Washington are the last word in mechanization. Hydraulically powered steel tables tilt up with apparent ease as wheat from the huge bulk trucks is dumped and stored in nearby elevators.

Courtesy, Washington Association Wheat Growers

Chapter IX

Wheat—Past, Present And Future...

IF BREAD IS THE STAFF OF LIFE, then it figures that wheat must be the staff of bread.

Wheat is the staff to a whole lot more than just bread though. Around it revolves a complex network of researchers, growers, handlers, millers, transporters and marketers. Economists, government regulatory agencies and legislative bodies also demand their fair share of comment in the world of wheat.

Millions owe their livelihood to wheat. Many more millions owe their lives to it.

Wheat is no longer the simple, unsophisticated plant it once was. Growing wild when first discovered by man thousands of years ago it is now the most analyzed, discussed, researched plant in existence.

The evolution of the wheat industry in the Northwest also coincides with the development of plant science and farm machinery. At first, farmers had only the simplest, almost primitive tools to work with. Then came the threshing machines, binders, headers, sweep powers, steam engines, gas tractors and combined-harvesters.

From small acreages the Pacific Coast grainmen went to huge plantings that had midwesterners scratching their heads in unbelief. Hillsides and flatlands came under the wheatman's plow.

Poking into the history of wheat in the Northwest reveals one fact—growing the golden kernels has been no soft touch for the farmers. Squeeze a wheat seed and it wouldn't be too surprising if it dripped with the sweat of thousands of farmers who have worked, prayed and nursed the grain into the Northwest's most valuable crop.

The wheat producer has bumped heads with every disease, every insect pest and every element of bad weather imaginable. Problems like these still persist today along with a price-cost squeeze of low prices and high costs which virtually bind a grainman in an economic strait jacket.

The result is that wheat means different things to different farmers. Some have gone broke in their attempts to grow the fickle grain. To the majority it has meant a reasonable living. To a few, wheat has brought a fortune.

The list of Northwest wheats which have topped the popularity list for a time and then faded is long. It includes such names as Turkey Red, Triplet, Fortyfold, Golden, Rex, Federation, Little Club, 128 Hybrid, Bluestem, Rio, Oro, Henry, Hymar, Marfed, Ridit, Jenkins Club, Foisey, Marquis, and Baart.

The varieties presently riding the crest of the wheat farmers' hit parade are Gaines and Nu

Gaines. A newcomer called Moro is making a strong bid for stardom.

In a few years the Gaines family of outstanding producers will undoubtedly be sidelined by even higher producing, more disease-resistant varieties. This is the name of the game. No line of wheat is so sacred it can't be replaced.

Weed, disease and pest control scientists along with plant breeders can never rest on their laurels. Smuts, rusts and insects keep bouncing back, creating new problems.

Wheat yields have gone from 10 to 15 bushels an acre on the east side of the mountains to 60 bushels and more at present. On the west side of the Cascades the early production hovered around 30 to 40 bushels. It has now accelerated up to 90 and 100 bushels or more to the acre.

Amazingly, this is just the beginning according to agronomy specialist Dr. Norman Goetze of the Oregon State University Cooperative Extension Service. He sees future yields of 200 bushels for west side wheat and 100 to 150 bushel yields for the bountiful grain east of the mountains.

This all revolves around continued improvements in weed control so plants have less competition for water, more effective use of fertilizers and new wheat varieties that yield more and can better withstand the attacks of fungus spores and hungry insects.

For the eastern sections it also depends a great deal on irrigation. It hardly sounds possible that the vast stretches of these intermountain wheatlands could be irrigated, but they can and will be. Many farmers in Lincoln county, Washington, have already installed costly sprinkler irrigation systems and have seen their production bound higher as a result. Farmers in other areas such as Arlington, Oregon, are in the process of analyzing the cost of getting water on their lands.

To some, the 200 bushel mark sounds as improbable today as the four minute mile did years ago. It doesn't take too deep a thinker though to realize that the 200 bushel goal will ultimately be cracked by the teamwork of farmers and scientists.

There has been great drama wrapped into the history of Northwest wheat. The only thing in the wheat industry that could possibly be more exciting than the past, is the future.

TURBINE TRACTOR. This is what International Harvester engineers suggest the near future turbine tractors will look like. The extra horsepower available from the turbine engine without extra weight will allow the tractor to plow, plant, and apply fertilizers, herbicides and insecticides in one pass—further reducing field compaction.

Courtesy, International Harvester Co.

AUTOMATIC TURBINES. Farther into the future—the 1980's to be exact—turbine tractors like this one will hit the field. The engine will be in the rear. A driver in the cab will be optional because the machine will be operated by a tape-control feed into the machine. The tractor could also respond to the furrow sensing device extending out from the machine shown in the lower portion of the picture. It will keep the tractor following a plowing pattern established on the first round. The farmer seated at the farmhouse, will be able to monitor robot machines working in fields miles away. He will also be able to shut down through a radioed signal in case of malfunction.

Courtesy, International Harvester Co.

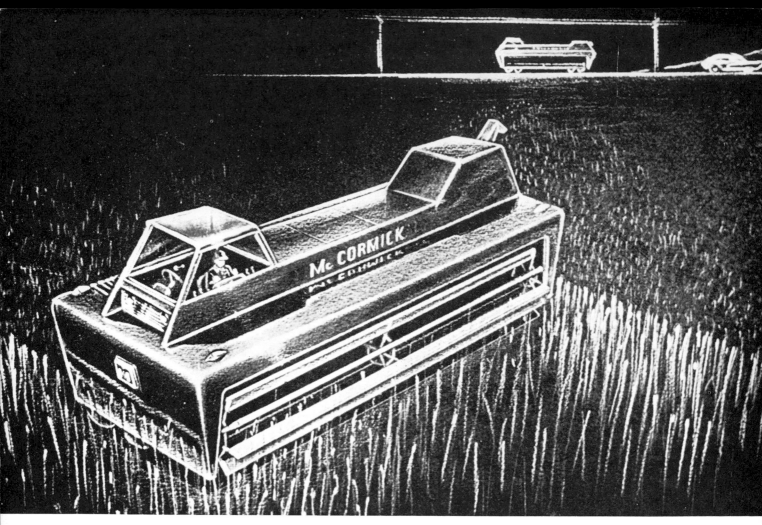

FUTURISTIC COMBINE. The combine of the 1980's will cut and harvest at about 12 miles per hour. Grain will be separated from the straw by centrifugal whirling chambers. Another method will be by putting an electrostatic charge to the kernels and separating them from the straw through the difference between the charges carried by the grain and the stalk. When the farmer finishes harvesting one field, he will pivot the wheels of the combine 90 degrees, turning his operator's seat and drive down the highway at the speed of a truck, taking up about the same space as a semitrailer.

Courtesy, International Harvester Co.

TRACTORS THAT DON'T LOOK LIKE TRACTORS. Another look at what tomorrow's tractors will look like came from the Ford study. The artist's conception shows an air-conditioned bubble-type cab that can be moved back over the equipment to check any problems.

Courtesy, Ford Tractor Co.

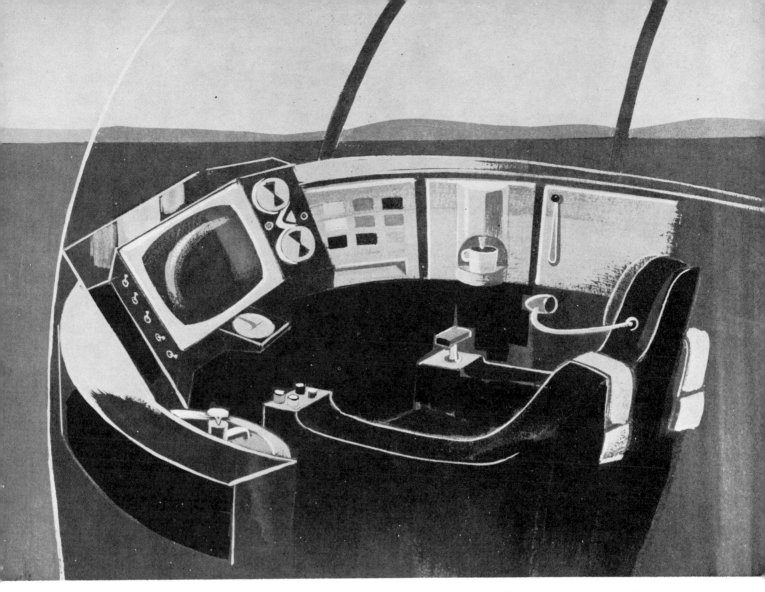

COMFORTABLE CAB. Predictions from Ford also indicate tractors of the future will be distant cousins of today's farm vehicles and their cabs will be just as advanced. This illustration depicts the cab of a tractor projected for the year 2000. Starting at the driver's right, and moving left, are a refrigerator, coffee maker, food warmer, a television set which is connected either to the farmer's headquarters or to other vehicles, there's even a sink. All controls are within arm's length of the driver so he can perform his jobs quickly and easily. This illustration is from "Agriculture 2000," a study conducted by Ford Motor Company's U. S. Tractor and Implement Operations to project the look of agriculture at the turn of the century.

Courtesy, Ford Tractor Co.

MAN AND WHEAT. The future of mankind is as closely linked to wheat as has been the past. As populations continue to explode, wheat will likely be the difference between a high and low standard of health—the difference between life and death for millions of people.

Courtesy, Washington Association of Wheat Growers

THE GOOD LIFE. Wheat farming, with all its advancements, still can't be considered an easy life. The family that makes its living by cropping wheat faces a number of challenges. The city dweller might wonder what these difficulties would be. They include heavy doses of hard work, small returns for time and money invested, a struggle against the elements of nature, and a sometimes undeserved poor public image. In spite of these hazards most farmers wouldn't trade their position for the life of a city dweller under any conditions. It's not an easy life, but it's a good life.

The Maynard Galbreath farm is located at the edge of Ritzville, Washington and looked like this in 1942.

Courtesy, A. M. Kendrick